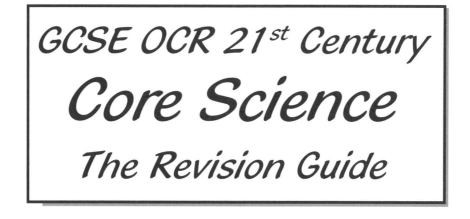

GCSE OCR 21st Century
Core Science
The Revision Guide

This book is for anyone doing **GCSE OCR 21st Century Core Science** at foundation level.

GCSE Science is all about **understanding how science works**.
And not only that — understanding it well enough to be able to **question**
what you hear on TV and read in the papers.

But you can't do that without a fair chunk of **background knowledge**. Hmm, tricky.

Happily this CGP book includes all the **science facts** you need to learn,
and shows you how they work in the **real world**. And in true CGP style,
we've explained it all as **clearly and concisely** as possible.

It's also got some daft bits in to try and make the whole
experience at least vaguely entertaining for you.

What CGP is all about

Our sole aim here at CGP is to produce the highest
quality books — carefully written, immaculately presented
and dangerously close to being funny.

Then we work our socks off to get them out to you — at the cheapest possible prices.

Contents

Published by Coordination Group Publications Ltd.

Editors:
Ellen Bowness, Tom Cain, Katherine Craig, Gemma Hallam, Sarah Hilton,
Kate Houghton, Andy Park, Rose Parkin, Kate Redmond, Rachel Selway,
Laurence Stamford, Jennifer Underwood, Julie Wakeling, Sarah Williams.

Contributors:
Mike Bossart, James Foster, Derek Harvey, John Myers, Andy Rankin,
Adrian Schmit, Moira Steven, Paul Warren, Andy Williams.

ISBN: 978 1 84146 625 5

With thanks to Glenn Rogers for the proofreading.
With thanks to Katie Steele for the copyright research.

Data used to construct table on page 85 from World Nuclear Association:
www.world-nuclear.org/info/inf05.htm

*With thanks to Science Photo Library for permission to reproduce the photographs used
on pages 21, 25, 56 and 71.*

With thanks to iStockphoto for permission to reproduce the photograph used on page 29.

Groovy website: www.cgpbooks.co.uk

Printed by Elanders Hindson Ltd, Newcastle upon Tyne.
Jolly bits of clipart from CorelDRAW®

The Scientific Process

This section <u>isn't</u> about how to 'do' science — but it does show you the way <u>most scientists</u> work, and how scientists try to find decent <u>explanations</u> for things that happen. It's pretty important stuff.

Scientists Come Up with Hypotheses...

1) Scientists try to <u>explain</u> things. Everything.

2) Scientists start by <u>observing</u> or <u>thinking about</u> something they don't understand. It could be anything, e.g. planets in the sky, a person suffering from an illness, what matter is made of... anything.

About 100 years ago, we thought atoms looked like this.

3) Then, using what they already know (plus a lot of creativity and insight), they work out an <u>explanation</u> (a <u>hypothesis</u>) that could explain what they've observed. Then they use their hypothesis to make a prediction that can be tested to provide further <u>evidence</u> to support the explanation. There are loads of people who can <u>collect data</u> perfectly well, but far fewer who can come up with a <u>decent explanation</u> that stands up to the next stage of the scientific process — <u>scrutiny</u> from other scientists.

...Then Look for Evidence to Test Those Hypotheses

1) A hypothesis is just a <u>theory</u> — a belief. And <u>believing</u> something is true doesn't make it true — not even if you're a scientist.

2) So the next step is to try and find <u>evidence</u> to support the hypothesis.

3) There are lots of different <u>types</u> of evidence:

- Results from <u>controlled experiments</u> in laboratories are great — a lab makes it easy to <u>control variables</u> so they're all kept constant (except the one you're investigating) — so it's easier to carry out a <u>fair test</u>.

- Data can also come from <u>studies</u>, e.g. of populations or samples of a population, or from <u>observations</u>, e.g. of the movement of planets, or of animal behaviour.

- The evidence might already be out there — the scientist might analyse <u>existing data</u> from a previous experiment to test the hypothesis.

- Rumours, hearsay and evidence from very small samples should all be taken with a <u>pinch of salt</u> — they're not very reliable.

Other Scientists Will Test the Hypotheses Too

1) Scientists report their findings to <u>other</u> scientists, e.g. by publishing their results in journals. Other scientists will use the evidence to make their <u>own predictions</u>, and they'll carry out their <u>own experiments</u>. (They'll also try to <u>reproduce</u> earlier results.) And if all the experiments back up a hypothesis, then scientists start to have a lot of <u>faith</u> in it, and accept it as a <u>theory</u>.

Then we thought they looked like this.

2) However, if a scientist somewhere in the world gets results that <u>don't</u> fit with the hypothesis, either those results or the hypothesis must be wrong.

3) This process of testing a hypothesis to destruction is a vital part of the scientific process. Without the '<u>healthy scepticism</u>' of scientists everywhere, we'd still believe the first theories that people came up with — like thunder being the belchings of an angered god (or whatever).

Science is a "real-world" subject...

Science isn't just about explaining things that people are curious about — if scientists can explain something that happens in the world, then maybe they can <u>predict</u> what will happen in the future, or even control <u>future events</u> — to make life a bit better in some way, either for themselves or for other people.

The Scientific Process

Evidence is the key to science — but not all evidence is equally good.
The way evidence is gathered can have a big effect on how trustworthy it is...

You Need Reliable Data, Not Opinion, to Justify an Explanation

1) The only scientific way to test a hypothesis is to gather appropriate data.
   (Please note — "It's true — I saw it on the telly" does not constitute appropriate data.)

2) And that data needs to be reliable — if it isn't, then it doesn't really help.

> RELIABLE means that the data can be reproduced by others in independent experiments or studies.

> In 1989 two scientists claimed that they'd produced 'cold fusion' (the energy source of the Sun — but without the enormous temperatures). It was huge news — if true, this could have meant energy from seawater — the ideal energy solution for the world... forever. However, other scientists just couldn't get the same results — i.e. the results weren't reliable. And until they are, 'cold fusion' isn't going to be generally accepted as fact.

Measurements aren't Always Accurate

You can't be sure that your measurements are accurate:

1) If you take a lot of measurements of the same thing, you won't always get the same result.

2) This might be because you're measuring lots of individual things, e.g. the heights of plants.
   Or the thing that you're measuring might not always be the same, e.g. the pollution level in the air.

3) It could also be that the thing stays the same but your measurement of it changes —
   e.g. your measuring equipment isn't very accurate, or you're not very good at measuring.

4) The best way to get a good estimate of the result is to repeat the measurement lots of times and take an average (see page 48 for more on this). The values that you get should all fall into a range that shows roughly where the real value is.

5) If you've got measurements that are obviously outside that range, then something might have gone wrong with those measurements.

If Evidence Supports a Hypothesis, It's Accepted — for Now

1) If pretty much every scientist in the world believes a hypothesis to be true because experiments back it up, then it usually goes in the textbooks for students to learn.

2) Our currently accepted theories are the ones that have survived this 'trial by evidence' — they've been tested many, many times over the years and survived (while the less good ones have been ditched).

3) However... they never, never become hard and fast, totally indisputable fact. You can never know... it'd only take one odd, totally inexplicable result, and the hypothesising and testing would start all over again.

4) There isn't a scientific answer to everything yet — not by a long way.

You expect me to believe that — then show me the evidence...

If a scientist thinks something's true, they need to produce evidence to convince others — it's part of testing a hypothesis. Along the way some hypotheses will be disproved — shown not to be true. But if no one can disprove a hypothesis, we've got a new explanation for something. It's how science works.

Correlation and Cause

Correlation and cause come up a lot in science. It's easy to get yourself into a mess of twisted logic, so read this page carefully and iron out anything you're not sure about right now.

A Correlation is a Relationship Between Two Factors

If there's a relationship between two things, then you can say there's a correlation between them, e.g.

- Sales of woolly hats increase when the weather is colder.
- There is a higher rate of lung cancer in smokers than non-smokers.
- More people who eat a diet high in saturated fat develop high blood pressure than those who eat a diet low in saturated fat.
- As pesticide use increases, the number of wild birds decreases.

A Correlation Doesn't Prove One Thing Causes Another

If you find a correlation between two things it's easy to think that one thing causes the other, but that's not always true — here's an example:

1) Primary school children with bigger feet tend to be better at maths. There's a correlation between the factor (big feet) and the outcome (better maths skills).

2) But it'd be crazy to say that having big feet causes you to be better at maths (and even weirder to say that being good at maths causes bigger feet...).

3) There's another (hidden) factor involved — their age.

4) Older children are usually better at maths. They also usually have bigger feet. Age affects both their maths skills and the size of their feet.

5) If you really thought that there was a link between shoe size and ability at maths, you'd test this by comparing children of the same age — you have to control the other factors, so that the only factor that varies is foot size.

So all correlation means is that there's a link between a factor and an outcome.

But even if there's a correlation between two factors, it doesn't always mean the outcome is inevitable. E.g. if you eat a diet high in saturated fat, it increases the chances that you'll get heart disease — it doesn't mean that you will get heart disease. You might stay fit and healthy all your life, and still be bopping away to Justin Timberlake at your 117th birthday party.

You've Got to do Valid Research

1) A scientist might hypothesise that there's a correlation between a factor and an outcome, and that one causes the other. To check their hypothesis, they should look for evidence by doing a scientific study.

2) They have to consider what other factors might influence the outcome, and minimise the effects of these other factors (see p.22 for more on this).

3) They also need a large enough sample for any correlation to be meaningful. A few isolated cases don't provide convincing evidence for or against a hypothesis, e.g. a study linking the MMR vaccine to autism used a sample of just 12 patients — many large studies since then suggest that there is no link.

All sheep die — Elvis died, so he must have been a sheep...

You read about correlations in the media all the time and reporters often make the mistake of thinking that if two things are correlated then one must cause the other. Get into the habit of questioning what you read or hear, and thinking about whether it's just people jumping to conclusions about the cause.

Risk

By reading this page you are agreeing to the <u>risk</u> of a paper cut or severe drowsiness that could affect your ability to operate heavy machinery... Think carefully — the choice is yours.

Nothing is Completely Risk-Free

1) <u>Everything</u> that you do has a <u>risk</u> attached to it.

2) Scientists often try to identify risks — they show <u>correlations</u> (see previous page) between certain activities (<u>risk factors</u>) and <u>negative outcomes</u>.

3) Some risks seem pretty <u>obvious</u>, or we've known about them for a while, like the risk of getting <u>heart disease</u> if you're overweight, or of having a <u>car accident</u> when you're travelling in a car.

4) As <u>new technology</u> develops it can bring new risks, e.g. some scientists believe that using a mobile phone a lot may be <u>harmful</u>. There are lots of risks we <u>don't know about</u> yet.

5) You can estimate the <u>size</u> of a risk based on <u>how many times</u> something has happened in a big sample (e.g. 100 000 people) over a given <u>period</u> (say, a year). The <u>more data</u> you have to base your assessment on, the more <u>accurate</u> your estimate is likely to be.

> So, to assess the risk of getting <u>struck by lightning</u> you could find out how many <u>people per 100 000</u> of the population is struck by lightning each year. That would give you an estimate of how likely it is that <u>you</u> will be struck during any year (as a number out of 100 000).
>
> Of course, that only actually tells you the risk for an <u>average</u> member of the population, which might not be an accurate estimate of an individual's risk. For example, if your favourite pastime is flying kites in thunderstorms whilst wearing full-plate steel armour, your risk is going to be higher than the average.

People Make Their Own Decisions About Risk

1) There's much more <u>chance</u> of cutting your finger during <u>half an hour of chopping veg</u> than of dying in a scuba-diving accident during <u>half an hour of diving</u>. People are usually happier to <u>accept</u> a high probability of an accident happening if the potential effects are <u>short-lived</u> (and fairly minor).

2) People are also more willing to accept risks if they get a significant <u>benefit</u> from the activity — e.g. car travel is quite <u>risky</u>, but the <u>convenience</u> of it means that people take the risk. Many people will also take risks for fun, e.g. in <u>extreme sports</u>.

3) <u>Freedom of choice</u> plays quite a big part, too. People tend to be more willing to accept a risk if they're <u>choosing</u> to do something, rather than if they're having the risk <u>imposed</u> on them.

We Have to Choose Acceptable Levels of Risk

1) People have to choose a <u>level</u> of risk that they find <u>acceptable</u>. This <u>varies</u> from person to person.

2) There are ways to reduce risks, but it's <u>impossible</u> to make anything completely safe. E.g. wearing a seatbelt reduces your risk of being killed in a car crash by about 50%.

3) <u>Governments</u> and scientists often have to <u>choose</u> levels of risk in various situations on behalf of <u>other people</u>. They'll often be <u>influenced</u> by <u>public opinion</u> though.

Take a risk — turn the page...

Risk isn't as <u>simple</u> as it looks — <u>risk management</u> is a really important job. It boils down to maths in the end though — it's all about the <u>probability</u> of something bad happening.

Science Has Limits

Science can give us amazing things — cures for diseases, space travel, heated toilet seats...
But science has its limitations — there are questions that it just can't answer.

Some Questions are Unanswered by Science — So Far

1) We don't understand everything. And we never will. We'll find out more, for sure — as more explanations are suggested and more experiments are done. But there'll always be stuff we don't know.

   For example, today we don't know as much as we'd like about climate change (global warming). Is climate change definitely happening? And to what extent is it caused by humans?

2) These are complicated questions. At the moment scientists don't all agree on the answers. But eventually, we probably will be able to answer these questions once and for all.

3) But by then there'll be loads of new questions to answer.

Other Questions are Unanswerable by Science

1) Then there's the other type... questions that all the experiments in the world won't help us answer — the "Should we be doing this at all?" type questions. There are always two sides...

2) Take embryo screening (which allows you to choose an embryo with particular characteristics). It's possible to do it — but does that mean we should?

3) Different people have different opinions. For example...

- Some people say it's good... couples whose existing child needs a bone marrow transplant, but who can't find a donor, will be able to have another child selected for its matching bone marrow. This would save the life of their first child — and if they want another child anyway... where's the harm?

- Other people say it's bad... they say it could have serious effects on the child. In the above example the new child might feel unwanted — thinking they were only brought into the world to help someone else. And would they have the right to refuse to donate their bone marrow (as anyone else would)?

4) This question of whether something is morally or ethically right or wrong can't be answered by more experiments — there is no "right" or "wrong" answer.

5) The best we can do is get a consensus from society — a judgement that most people are more or less happy to live by. Science can provide more information to help people make this judgement, and the judgement might change over time. But in the end it's up to people and their conscience.

There will be Different Benefits and Costs for Different People

Any scientific development will have benefits and costs for different groups of people.

- Some people argue that the right solution to an ethical problem is the one that leads to the best outcome for the majority of people involved. E.g. the results from embryonic stem cell research could be used to cure millions of people suffering from life-threatening diseases. It can be argued that it's worth the loss of the potential life of the embryo to improve the lives of so many.

- Another argument is that some things are just wrong or unnatural, and can never be justified. E.g. embryonic stem cell research is seen by some people as 'playing God'. Do we have the right to choose who lives and who dies?

Science doesn't have all the answers...

Nothing's ever simple, and science can only help you so far. It can tell you what's possible, but people have to decide for themselves whether or not it's ethically or environmentally acceptable. And some people have extreme views on what we should or shouldn't do so it can be hard to make decisions.

Genes, Chromosomes and DNA

Welcome to the first Biology bit of the OCR 21st Century Science Revision Guide. You're gonna love it. If you ask me there's no better way to get the ball rolling than with a bit of good old genetics.

The Nucleus Contains Instructions for Development

The cell nucleus contains genes — the instructions responsible for the development of a freshly fertilised egg into a fully grown human being.

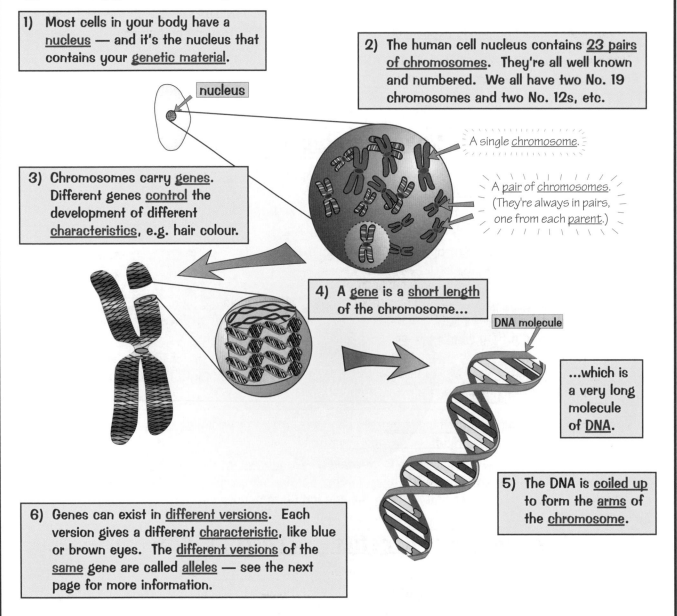

1) Most cells in your body have a nucleus — and it's the nucleus that contains your genetic material.

2) The human cell nucleus contains 23 pairs of chromosomes. They're all well known and numbered. We all have two No. 19 chromosomes and two No. 12s, etc.

nucleus

A single chromosome.

A pair of chromosomes. (They're always in pairs, one from each parent.)

3) Chromosomes carry genes. Different genes control the development of different characteristics, e.g. hair colour.

4) A gene is a short length of the chromosome...

DNA molecule

...which is a very long molecule of DNA.

5) The DNA is coiled up to form the arms of the chromosome.

6) Genes can exist in different versions. Each version gives a different characteristic, like blue or brown eyes. The different versions of the same gene are called alleles — see the next page for more information.

Each gene is an instruction for the cell about how to make a certain protein. Having different versions of proteins means that we end up with different characteristics.

It's hard being a DNA molecule, there's so much to remember...

This is the top and bottom of genetics, so you definitely need to understand everything on this page or you'll find the rest of this topic dead hard. The best way to get all of these important facts engraved in your mind is to visit your local stone mason. Alternatively, cover the page, scribble down the main points and sketch out the diagrams... see how much you've remembered, then learn the bits you missed.

Inheritance

Thought that being cursed with your dad's <u>huge nose</u> was just a cruel coincidence?
Well, there's a bit more to it than that...

Children Resemble Both Parents...

The genes in your cells are a <u>mixture</u> of your parents' genes. You inherit <u>half</u> from your <u>mum</u> and <u>half</u> from your <u>dad</u> due to <u>sexual reproduction</u>:

1) Sexual reproduction is when a <u>sperm fertilises an egg</u> to create an embryo (oh, the romance of it all).

2) The <u>sex cells</u> (the sperm and the egg) are different from ordinary body cells — they contain <u>23 single chromosomes</u>, instead of 23 pairs.

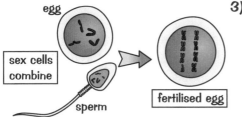

egg

sex cells combine

fertilised egg

sperm

3) When the sperm <u>fertilises</u> the egg, the chromosomes in the sperm combine with the chromosomes in the egg — the fertilised egg then has 23 <u>pairs</u> of chromosomes, just like an ordinary body cell.

4) So, one chromosome in every <u>pair</u> has come from <u>each parent</u> (remember — humans have 23 pairs of chromosomes in every body cell).

5) The two chromosomes in a pair carry the <u>same genes</u>, in the <u>same places</u>. Because the two chromosomes in a pair came from <u>different parents</u>, they might have different versions of these genes — called <u>alleles</u>.

6) So, most children look a bit like <u>both</u> of their parents — this is because they get <u>some</u> of their alleles from each of their parents.

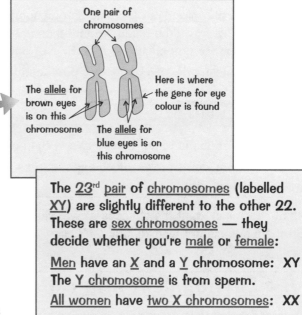

One pair of chromosomes

The <u>allele</u> for brown eyes is on this chromosome

Here is where the gene for eye colour is found

The <u>allele</u> for blue eyes is on this chromosome

The <u>23rd</u> <u>pair</u> of <u>chromosomes</u> (labelled <u>XY</u>) are slightly different to the other 22.
These are <u>sex chromosomes</u> — they decide whether you're <u>male</u> or <u>female</u>:

<u>Men</u> have an <u>X</u> and a <u>Y</u> chromosome: XY
The <u>Y</u> chromosome is from sperm.
<u>All women</u> have <u>two X</u> chromosomes: XX

...But aren't Identical to Either

They won't be <u>exactly</u> like either one of their parents because they haven't got <u>all</u> the same alleles — half came from the <u>other</u> parent.

Children Resemble Siblings but aren't Identical

You may wonder then, why all people don't look identical to their <u>brothers</u> and <u>sisters</u>? After all, they inherited half of their genes from the <u>same mum</u> and the other half from the <u>same dad</u>.

1) It's all down to how <u>sex cells are made</u> and <u>how they combine</u> in a random fashion (see p.8). There are millions of <u>different combinations</u> possible.

2) In fact, every child will have a new, <u>unique</u>, combination of alleles — that's why no two people in the world are exactly the same (apart from identical twins, but you don't need to worry about that till p.10).

Genes — they always come in pairs...

I wish I inherited <u>more</u> from my parents — all I got from them was my wonky ears, one bushy eyebrow and a grandfather clock... I would have preferred a <u>mansion</u> to be honest, but you can't win 'em all.

Variation

So, organisms of the same species (e.g. you and your sibling) will usually look slightly different from each other. These differences are called the variation within a species — and there are two things that create variation: genes and environment.

Sexual Reproduction Creates Genetic Variation

Sexual reproduction creates variation in genes (and so variation in the characteristics you inherit). This happens when sex cells are formed and at fertilisation.

1) SEX CELL FORMATION

Sex cells are formed from body cells, which have 23 pairs of chromosomes (don't forget that the two chromosomes in a pair are never identical because they have different alleles).

1) When the sex cells are formed the pairs of chromosomes in the body cell are separated and one chromosome from each pair goes into one sex cell and one goes into another sex cell.

2) This means that each sex cell gets chromosomes with different alleles.

3) There are millions of possible chromosome combinations that can be produced from the separation of 23 pairs. (You could get chromosome 1, 2, 4, 6... of one version and 3, 5, 7 of the other version, or you could get 1, 3, 5, 7 of one version and 2, 4, 6 of another etc.)

4) This means that every sex cell produced by one individual will be genetically different.

2) FERTILISATION

1) Fertilisation is when the sperm and the egg join to form a new cell with 23 pairs of chromosomes.

2) When a woman releases an egg it can be fertilised by any one of millions of different sperm released by her partner.

All of this means that the chances of two siblings being identical are absolutely minuscule. Brothers and sisters tend to look a bit alike, but there are always differences.

The Environment Also Causes Variation in Characteristics

1) Most variation in organisms (including humans) is caused by a mixture of genetic and environmental factors.

2) Almost every single aspect of an organism is affected by the environment in some way, however small. In fact it's a lot easier to list the factors which aren't affected in any way by environment — e.g. eye colour, blood group and inherited disorders (like Huntington's disorder and cystic fibrosis).

If you're not sure what 'environment' means, think of it as 'upbringing' instead.

3) Pretty much everything else is affected by the environment. Here's an example:

Some people are more likely to get certain diseases (e.g. cancer and heart disease) because of their genes. But lifestyle also affects the risk, e.g. if you smoke or only eat junk food then you're more likely to get ill.

Revision? I'm relying on my genetic ability...

Somehow, I doubt that you've inherited an encyclopedic knowledge of OCR 21st Century Science from your parents, so you'd best hedge your bets and do some revision. Whilst you're at it make sure that the environmental factors don't let you down — you'll need plenty of coffee and chocolate.

Single Gene Inheritance

In genetics you're never more than a stone's throw away from a genetic diagram. They show the inheritance of a single gene (which causes one characteristic, e.g. blood group). Most characteristics are actually determined by loads of genes working together, e.g. height.

Genetic Diagrams Show the Possible Alleles of Offspring

1) As I keep saying, alleles are different versions of the same gene.

2) Most of the time you have two of each gene (i.e. two alleles) — one from each parent.

3) If the alleles are different you have instructions for two different versions of a characteristic (e.g. blue eyes or brown eyes), but you only develop one version of the two (e.g. brown eyes). The version of the characteristic that appears is caused by the dominant allele. The other allele is said to be recessive.

4) In genetic diagrams, letters are used to represent alleles. Alleles that produce dominant characteristics are always shown with a capital letter, and alleles that produce recessive characteristics with a small letter.

5) You can have two alleles the same for a particular gene, e.g. CC or cc. Or two different alleles for a gene, e.g. Cc.

You Need to Interpret Genetic Diagrams

Imagine you're cross-breeding hamsters, and that some have a normal, boring disposition while others have a leaning towards crazy acrobatics. And suppose you know the behaviour is due to one gene...

Let's say that the allele which causes the crazy nature is recessive — so use a 'b'.
And normal (boring) behaviour is due to a dominant allele — call it 'B'.

1) For an organism to display a recessive characteristic, both its alleles must be recessive — so a crazy hamster must have the alleles 'bb'.

2) However, a normal hamster could be BB or Bb, because the dominant allele (B) overrules the recessive one (b).

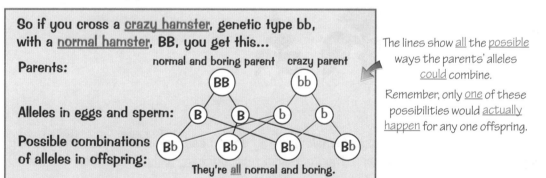

So if you cross a crazy hamster, genetic type bb, with a normal hamster, BB, you get this...

Parents: normal and boring parent   crazy parent
BB   bb

Alleles in eggs and sperm: B   B   b   b

Possible combinations of alleles in offspring: Bb   Bb   Bb   Bb

They're all normal and boring.

The lines show all the possible ways the parents' alleles could combine.

Remember, only one of these possibilities would actually happen for any one offspring.

If two of these offspring now breed they will produce new combinations of alleles in their kids:

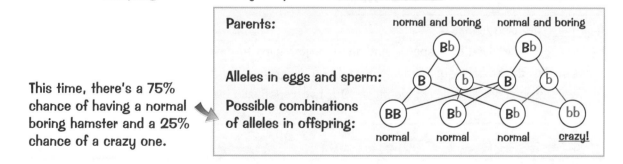

Parents: normal and boring   normal and boring
Bb   Bb

Alleles in eggs and sperm: B   b   B   b

Possible combinations of alleles in offspring: BB   Bb   Bb   bb
normal   normal   normal   crazy!

This time, there's a 75% chance of having a normal boring hamster and a 25% chance of a crazy one.

It's not just hamsters that have the wild and scratty allele...

...my sister definitely has it too. Remember, 'results' like this are only probabilities. It doesn't mean it'll actually happen. (Most likely, you'll end up trying to contain a mini-riot of nine lunatic baby hamsters.)

Clones

Cloning seems to be a hot topic in the news at the moment, but nature has been making clones for millions of years and nobody seems that bothered.

Clones **are** Genetically Identical **Organisms**

1) Sexual reproduction produces offspring that are genetically different. But other methods of reproduction can produce offspring that are genetically identical to each other.

2) Genetically identical organisms are called clones. Clones have the same genes, and also the same alleles of those genes.

3) Because clones have the same alleles, any differences between them must be due to differences in their environment — for example, the amount of food available.

Clones **Can be Produced** Naturally...

(i) By Asexual Reproduction

Some organisms reproduce asexually (without sexual reproduction). This means that there is only one parent, and the offspring are genetically the same as each other and the parent.

1) Most bacteria reproduce like this — they simply divide into two.

2) Many plants can also reproduce asexually. They produce an offshoot, which develops as a separate plant.

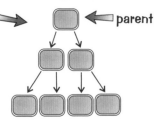

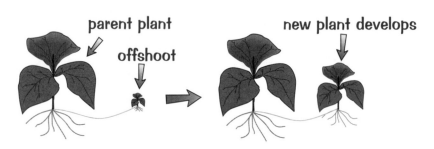

parent plant offshoot new plant develops

3) A few animals can reproduce asexually. Female greenfly don't need to mate — they can just lay eggs, which develop into more females. They can also reproduce sexually, when they feel like it.

(ii) When Cells of an Embryo Split

Identical twins are also clones.

1) A single egg is fertilised by a sperm, and an embryo begins to develop as normal.

2) Occasionally, the embryo splits into two, and two separate embryos begin to develop.

3) The two embryos are genetically identical. So, two genetically identical babies are born.

Attack of the clones — panic over, it's only a few plants...

Remember, not all twins are identical (remember Schwarzenegger and DeVito). Non-identical twins are not clones — genetically speaking they're just as different as normal siblings would be. Anyway, you need to learn everything on the page, cover it up and see if you can clone it onto a blank piece of paper.

Genetic Disorders

Unfortunately, some genes can cause nasty disorders.

Genetic Disorders are Caused by Faulty Alleles

1) Some <u>disorders</u> are <u>inherited</u> — one or both parents carry a <u>faulty allele</u> and pass it on to their children.
2) <u>Cystic fibrosis</u> and <u>Huntington's disorder</u> are both caused by a faulty allele of a <u>single gene</u>.

Some Genetic Disorders are Caused by Recessive Alleles...

<u>Defective alleles</u> are responsible for <u>genetic disorders</u>. Most of these defective alleles are <u>recessive</u>.

<u>Cystic fibrosis</u> is a <u>genetic disorder</u> of the <u>cell membranes</u>. It <u>results</u> in the body producing a lot of thick, sticky <u>mucus</u> in the <u>air passages</u>, <u>gut</u> and <u>pancreas</u>, causing <u>breathing</u> and <u>digestive</u> problems.

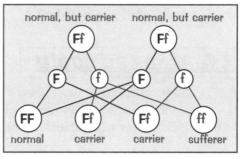

1) The allele which causes cystic fibrosis is a <u>recessive allele</u>, 'f', carried by about <u>1 person in 25</u>.
2) Because it's recessive, people with only <u>one copy</u> of the allele <u>won't</u> have the disorder — they're known as <u>carriers</u>.
3) For a child to have a chance of inheriting the disorder, <u>both parents</u> must be either <u>carriers</u> or <u>sufferers</u>.
4) As the diagram shows, there's a <u>1 in 4 chance</u> of a child having the disorder if <u>both</u> parents are <u>carriers</u>.

Knowing how inheritance works can help you to interpret a <u>family tree</u> — this is one for <u>cystic fibrosis</u>:

There's a <u>25%</u> chance that the new baby will be a sufferer and a <u>50%</u> chance that it will be a carrier because both of its parents are carriers. The case of the new baby is just the same as in the genetic diagram above — so the baby could be <u>normal</u> (FF), a <u>carrier</u> (Ff) or a <u>sufferer</u> (ff).

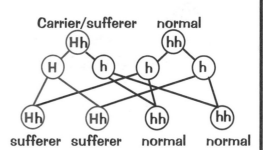

...Others are Caused by Dominant Alleles

1) Unlike cystic fibrosis, <u>Huntington's disorder</u> is caused by a <u>dominant</u> allele.
2) The disorder causes <u>shaking</u>, <u>erratic body movements</u> and <u>mental deterioration</u> — and there's <u>no cure</u>.
3) The dominant allele means there's a <u>50%</u> chance of each child inheriting the disorder if just one parent is a carrier. These are seriously grim odds.
4) The 'carrier' parent will of course be a <u>sufferer</u> too since the allele is dominant, but the symptoms do not appear until after the age of 40, by which time the allele has been <u>passed on</u> to children and even grandchildren. Hence the disorder persists.

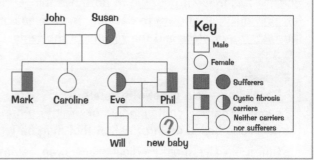

So, if one parent is a sufferer, there's a 1 in 2 chance of each of their children having the disorder.

Unintentional mooning — caused by faulty genes...

We <u>all</u> have defective genes in us somewhere — but usually they don't cause us a problem (as they're often <u>recessive</u>, so if you have a healthy <u>dominant</u> allele too, you'll be fine). Phew-ee.

Genetic Testing

Nowadays it's possible to test for all sorts of different genetic conditions, but this has thrown up some pretty big questions about what we should and shouldn't be doing.

Embryos and Fetuses Can be Screened for Some Disorders...

1) When embryos are produced using IVF (in vitro fertilisation or test-tube babies), doctors can test the embryos to check if they've got certain genetic disorders. This is called genetic screening. This is especially important if there's concern that one of the parents might carry alleles for a genetic disorder.

2) Doctors produce more embryos than they need during IVF. The embryos are screened and only healthy ones are chosen to be implanted into the mother's womb.

3) Doctors can also test fetuses in the womb by testing the fluid surrounding the fetus — this is called amniocentesis.

...And So Can Adults

Adults can be checked to see if they carry alleles for genetic disorders, e.g. Huntington's disorder. If a couple find out that their children might inherit a genetic disorder, then they might decide not to have children (or might decide to have their embryos / fetus screened).

There are Issues Surrounding Genetic Testing

Science has made it possible to do these things, but science can't say whether we should do them. That's for individuals or society to decide. The issues surrounding genetic testing fall into two general categories: the tests themselves and the results of the tests.

Tests

Reliability — No genetic tests are 100% accurate. There are always some errors due to things like samples getting contaminated or misinterpretation of results. This means people have to make decisions based on information that may be incorrect.

Safety — Like most medical procedures, tests carried out during pregnancy aren't 100% safe, e.g. amniocentesis causes a miscarriage in 0.5 to 1% of cases.

Results

When tests are carried out, there can be more decisions to make once the results have arrived. E.g.

1) If a test result is positive, should other members of a family be tested? Some people may prefer not to know, but is this fair on any partners or future children they might have?

2) Is it right for someone who's at risk of passing on a genetic condition to have children? Is it fair to put them under pressure not to, if they decide they want children?

3) If a test carried out during a pregnancy is positive, is it right to terminate the pregnancy? Perhaps the parents wouldn't be able to cope with a sick or disabled child, but does that child have less right to life than a healthy child? Some people think abortion is always wrong, whatever the circumstances.

Get ready for some genetic testing in the exam...

This is one of the really big topics so chances are it'll crop up somewhere in one of your exams. You need to understand what genetic testing is and when it's used. You also need to be clued up on the ethical issues — be aware of all the pros and cons, not just your own opinion.

Gene Therapy

Gene therapy is a new science with exciting possibilities. There are hopes that it could really help people with genetic disorders like cystic fibrosis. But, like other areas of genetics, there are concerns that it may lead to some pretty serious problems...

Gene Therapy Could Improve Lives...

Gene therapy may soon make it possible to treat or prevent genetic disorders. It works by inserting a healthy copy of a gene to correct a faulty gene.

Example 1 — Treating Cystic Fibrosis

1) Cystic fibrosis (CF) affects about 1 in 2500 people in the UK.
2) Scientists are trying to treat cystic fibrosis with gene therapy. One method being tried out is to use a virus to insert a healthy copy of the gene into cells in the airways.
3) There are some problems — for example, at the moment the effect wears off after a few days. But there are big hopes that gene therapy will one day mean CF can be treated effectively.

Example 2 — Preventing Breast Cancer

1) Breast cancer is a disease where cells grow uncontrollably, causing tumours. Some people inherit defective genes that predispose them to getting breast cancer.
2) In theory it's possible to supplement the faulty genes with working, healthy ones using gene therapy.
3) Scientists can't do this yet because they need to identify the genes involved (there's more than one), and they need to target the working genes to the breast area.
4) But not all breast cancers are caused by faulty genes — so giving someone the working gene will mean they're no longer predisposed, but they could still get breast cancer anyway.

...But There Could be Problems

Being able to prevent and treat diseases sounds great, but there are some potential drawbacks:

1) Gene therapies carry risks themselves — they might affect other cells besides the target cells. There are concerns that they could even cause cancers instead of preventing them.
2) Some people have concerns about 'playing God', and meddling with things that should be left well alone.

Gene therapy — talk things through with your Levi's...

Gene therapy is another ethical minefield. It throws up all sorts of questions, including exactly what should be classed as a 'disease' that needs fixing. Many people would argue that treating cystic fibrosis is a good thing, but what about high blood pressure, or an inability to tan... where does it stop?

Stem Cells

Stem cell research has <u>exciting possibilities</u> too, but it's also pretty <u>controversial</u>.

Stem Cells *Can Become ANY Type of Cell*

1) Most cells in your body are <u>specialised</u> for a particular job. E.g. white blood cells are brilliant at fighting invaders, but they can't carry oxygen like red blood cells.

2) <u>Differentiation</u> is the process by which a cell <u>changes</u> to become <u>specialised</u> for its job. In most <u>animal</u> cells, the ability to differentiate is <u>lost</u> at an early stage.

3) Some cells are <u>undifferentiated</u>. They can develop into <u>different types of cell</u> depending on what <u>instructions</u> they're given. These cells are called <u>STEM CELLS</u>.

4) Stem cells are found in early human <u>embryos</u>.

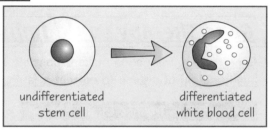
undifferentiated stem cell → differentiated white blood cell

Stem Cells *May be Able to Cure Many Diseases*

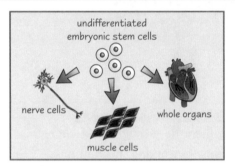
undifferentiated embryonic stem cells → nerve cells, whole organs, muscle cells

Stem cells are <u>exciting</u> to doctors and medical researchers because they have the potential to turn into <u>any</u> kind of cell at all, which means they can be used to produce any new tissue.

1) Medicine <u>already</u> uses adult stem cells to cure <u>disease</u>. For example, people with some <u>blood diseases</u> (e.g. <u>sickle cell anaemia</u>) can be treated by <u>bone marrow transplants</u>. Bone marrow contains <u>adult stem cells</u> that can turn into <u>new blood cells</u> to replace the faulty old ones.

2) Scientists can also <u>extract</u> stem cells from very early human <u>embryos</u> and <u>grow</u> them.

3) These embryonic stem cells could be used to <u>replace faulty cells</u> in sick people — you could make <u>heart muscle cells</u> for people with <u>heart disease</u>, <u>insulin-producing cells</u> for people with <u>diabetes</u>, <u>nerve cells</u> for people <u>paralysed by spinal injuries</u>, and so on.

4) To get cultures of <u>one specific type</u> of cell, researchers try to <u>control</u> the differentiation of stem cells by changing the environment they're growing in. It's still a bit hit and miss — lots more <u>research</u> is needed.

The Use of Stem Cells is *Controversial*

There are lots of arguments <u>for</u> and <u>against</u> the use of stem cells from <u>embryos</u>. The main argument is basically between those who feel that the <u>benefits to ill people</u> outweigh the fact that an <u>embryo</u> has to be <u>destroyed</u> and those who <u>disagree</u> with 'killing' an <u>embryo</u>.

In favour of using stem cells from embryos	Against using stem cells from embryos
Early human embryos are only balls of cells — they're not yet human beings with rights.	These new 'cures' are still unproven. Research may be carried out which turns out to be useless, causing the unnecessary deaths of many embryos.
The needs of an adult or a child with a crippling disease are more important than the needs of an unborn embryo.	It's especially important to protect the rights of the unborn, because they can't speak up for themselves.
Research using a few embryos might bring benefits to thousands of people.	Destroying human life is wrong, whatever the circumstances.
The embryos used are often left over from IVF so would be destroyed anyway.	Only God should have the power to create and destroy life.

My brother played God in a school play... I was Mary...

Another exciting prospect is making an <u>embryo</u> that's a <u>clone</u> of a patient with an illness. Stem cells could then be extracted from the embryo — the big advantage is that the body <u>wouldn't reject</u> them.

Science and Ethics

Genetics can be even more controversial than when Harold Bishop came back from the dead in Neighbours.

Scientific Advances Usually Have Benefits and Costs

Scientists have created loads of new technologies that could improve our lives.
In biotechnology, for example, benefits include:

1) Being able to identify whether a fetus has a genetic disorder.
2) Identifying whether an adult will get a disease, such as Huntington's, later in life.
3) Growth of human stem cells to be used as spare parts, for example in transplant surgery.
4) The possibility of being able to insert healthy alleles into the cells of people with genetic disorders.

However, it's not all good news. Some of the costs are:

1) Scientific research is expensive and uncertain. Perhaps the money that goes into research would be better spent on things like building new hospitals and training doctors.
2) No one knows what the results of genetic experiments will be, which is risky. Who knows what might happen if scientists alter the pattern of our DNA?
3) Some people think that using stem cells from human embryos is an unacceptable cost.

There are many different opinions about whether or not the benefits outweigh the costs...

There are Two Key Approaches to Ethical Dilemmas

Of course, there are many different ways of looking at ethical dilemmas.
For the exam you need to be able to develop arguments based on two key principles:

1) You may think that certain actions are always unnatural or wrong. This means that, whatever the possible benefits, you feel these actions are unacceptable. Some people would say that research on human embryos comes into this category.

2) You may also say that the right decision is the one that brings the greatest benefit to the greatest number of people. Some people would argue that embryo research does more good than harm and so it's acceptable.

Another consideration is that some people think it's unfair for people to benefit from something which is only possible because someone else has suffered or taken a risk. The trouble is, all new medicines and medical procedures must be tested on human volunteers — and that always involves some risk.

The Law is Sometimes Involved Too

As well as individuals, the law is involved in regulating scientific research.

1) Animal research is regulated. For example, in the UK scientists researching on vertebrates must have a licence and they must show that the likely benefits of the research outweigh the animals' suffering.
2) In the UK, no research is allowed on embryos older than 14 days.
3) Genetic manipulation is also regulated. In Britain, genetic manipulation of human body cells is allowed, but the modification of reproductive cells (sperm and egg cells) isn't.
4) There are regulations about the effect of research on the environment.

Hmmm, tricky...

As you can see, this unit isn't just about knowing your facts — you need to think about the ethical issues in genetics and the kinds of arguments people consider when making decisions about what should be done.

Revision Summary for Module B1

Well done, you've made it to the end of the first section of the book — only eight more to go. I know you're chomping at the bit to move on to C1, but you ain't quite finished here yet. Have a go at all of these questions and see how many you can get right. Then go back through the section and learn the bits you got stuck on. If you can do all of them without looking then you'll be onto a winner for the exam.

1) Where is DNA found within a cell?

2) Are chromosomes usually found on their own, in pairs or in threes?

3) What is a gene?

4) Why are genes so important?

5) What are alleles?

6) Name the two types of sex cell.

7) Why do most children look a bit like both of their parents but not identical to either?

8) What's the difference between males and females in terms of their chromosomes?

9) How does sexual reproduction create variation?

10) Name a characteristic that is determined by genes alone.

11) Give one example of how the environment affects variation.

12) Is height determined by genes, the environment or both?

13)* Here's a genetic diagram for the possible inheritance of an allele for loving Lemar albums. The allele is dominant and one parent is AA and the other is Aa.

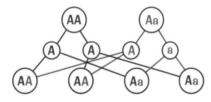

a) What is the chance of the offspring loving Lemar albums?

b) What is the chance of the offspring not loving Lemar albums?

14) Write a definition of the term 'clone'.

15) How do plants reproduce asexually?

16) How are identical twins formed?

17) Is the allele for cystic fibrosis dominant or recessive?

18) What are the symptoms of cystic fibrosis?

19) What are the symptoms of Huntington's disorder?

20) What is the chance of a child inheriting Huntington's disorder if one of their parents is a carrier?

21) Why are embryos screened during IVF treatment?

22) What are the two main problems with genetic screening?

23) Name two diseases that may be treated in the future using gene therapy.

24) What is differentiation?

25) List some ethical arguments for and against using stem cells from embryos.

26) Give some examples of laws that regulate scientific research.

* Answers on page 100.

The Atmosphere

We take air for granted — breathing it in and out and never giving it a second thought. But have a think about it — loads of things that we do every day are polluting and damaging our atmosphere.

The Atmosphere is a Mixture of Gases

1) The Earth is surrounded by an <u>atmosphere</u> — a mixture of <u>gases</u>.

2) The atmosphere is mostly made up of nitrogen, oxygen and argon:

> Nitrogen 78%   Oxygen 21%   Argon 1%

This blue haze is the Earth's <u>atmosphere</u>.

3) The figures above are rounded slightly — the atmosphere also contains small amounts of <u>carbon dioxide</u> and various <u>other gases</u>, and varying amounts of <u>water vapour</u>.

Human Activity is Changing the Atmosphere

1) The concentrations of <u>nitrogen</u>, <u>oxygen</u> and <u>argon</u> in the atmosphere are pretty much <u>constant</u>.

2) But... human activity is <u>adding</u> small amounts of <u>pollutants</u> to the air. There are five pollutants you need to know about:

> A <u>pollutant</u> is a chemical that's harmful because it's in the 'wrong' place.

- carbon dioxide (see p.19)
- carbon monoxide (see p.19)
- particles of carbon (see p.19)
- sulfur dioxide (see p.20)
- nitrogen oxides (see p.20)

3) These pollutants come from many different sources. You need to know about pollution from <u>burning fuels</u> — in <u>power stations</u> and <u>vehicles</u>. (See pages 18 to 20.)

4) Some pollutant gases are <u>directly harmful</u> to humans — they can cause <u>disease</u> or <u>death</u> in people who breathe in large enough quantities. E.g. vehicle exhausts contain pollutants which contribute to breathing problems like <u>asthma</u>.

5) Pollutants can also harm us <u>indirectly</u>, by damaging our <u>environment</u>. For example:

- Some pollutants cause <u>acid rain</u>, which pollutes rivers and lakes — killing the <u>fish</u> which people catch and <u>eat</u>.
- Other pollutants are thought to be leading to <u>climate change</u> (see p.59), which could bring <u>all sorts of problems</u> — rising sea levels, disruption to farming, more hurricanes...

Most Fuels are Hydrocarbons

> A <u>compound</u> is where different atoms are <u>bonded together</u> chemically.

Most of the fuels we burn in cars, trains, planes etc. are <u>hydrocarbons</u>.

1) A hydrocarbon is a <u>compound</u> containing <u>hydrogen</u> and <u>carbon</u> atoms only.

2) Fuels such as <u>petrol</u>, <u>diesel fuel</u> and <u>fuel oil</u> are mixtures of <u>hydrocarbons</u>.

3) Many <u>power stations</u> also burn hydrocarbons, e.g. natural gas.

Many other power stations burn <u>coal</u>. Coal <u>isn't</u> a hydrocarbon — it's <u>just carbon</u> (with a few impurities).

Polluting the atmosphere — smoking at parties...

The atmosphere's pretty important — without it, there'd be no <u>fresh air</u> to breathe. In some places though, the air isn't very 'fresh' at all. It's said that living in Mexico City does the same damage to your health as smoking 40 cigarettes a day — the city's built in a <u>dip</u>, so traffic fumes tend to <u>hang over</u> it.

Chemical Reactions

If you <u>burn</u> a fuel, a chemical reaction occurs — atoms from the <u>fuel</u> react with atoms from the <u>air</u> — and the atoms <u>rearrange themselves</u> to make <u>other substances</u>.

Chemical Reactions *Happen When Atoms are Rearranged*

<u>Burning</u> is a type of <u>chemical reaction</u>. Almost all chemical reactions involve atoms <u>changing places</u>.

1) When a <u>hydrocarbon</u> burns, the <u>hydrogen</u> atoms in the fuel combine with <u>oxygen</u> atoms from the air to make <u>hydrogen oxide</u> (otherwise known as <u>water</u>)...

2) ... and the <u>carbon</u> atoms in the fuel combine with <u>oxygen</u> from the air to make <u>carbon dioxide</u>.

Burning a Hydrocarbon — Example

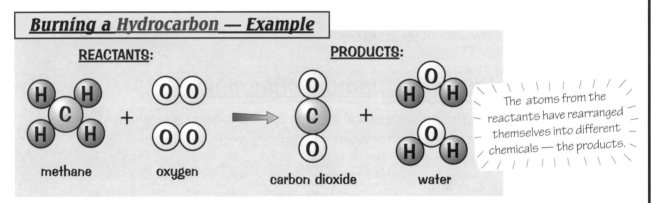

REACTANTS: PRODUCTS:

methane   oxygen   carbon dioxide   water

The atoms from the reactants have rearranged themselves into different chemicals — the products.

3) In the reaction shown above, the <u>reactants</u> are methane (a hydrocarbon) and oxygen...

4) ...and the <u>products</u> are carbon dioxide and water.

5) <u>No atoms 'disappear'</u> during the reaction — if you count up all the atoms in the reactants (1 carbon, 4 hydrogens and 4 oxygens), you'll find they're all still there in the products.

6) It's the same with <u>any</u> chemical reaction — the atoms get <u>shuffled about</u>, but they're <u>all still there</u>.

7) When <u>coal</u> burns, you mostly get carbon dioxide.

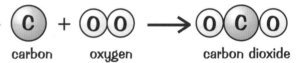

carbon   oxygen   carbon dioxide

Whenever we burn fuels containing carbon (like coal, petrol, diesel, etc.) we <u>add</u> <u>carbon dioxide</u> to the atmosphere. But that's not <u>all</u> we add — fuels usually contain various <u>impurities</u>, which also combine with oxygen and make <u>other pollutant gases</u>.

Reactants and Products Often Have Very Different Properties

1) When atoms <u>rearrange</u> themselves in a reaction, the products that are formed have <u>their own properties</u> — which can be <u>very different</u> from the properties of the reactants.

2) For example, <u>carbon</u> is a black solid and <u>oxygen</u> is a colourless gas (at room temperature). When they react together, as <u>coal burns</u>, the product is <u>carbon dioxide</u> — which is a <u>colourless gas</u>, but with very different properties from oxygen (it's heavier for a start, and can be toxic).

The burning question is — do you know it all...

Chemical reactions are great — atoms rearrange themselves and you get totally different stuff from the stuff you started with. For example, <u>sodium</u> is a metal that will burn you if you touch it, and <u>chlorine</u> is a <u>poisonous</u> gas. But when they react together, you get <u>sodium chloride</u> — nice, edible <u>table salt</u>.

Air Pollution — Carbon

Coal, petrol, diesel, etc. are <u>fossil fuels</u> — they were formed from the remains of dead plants and animals.

Different Forms of Carbon Pollution Cause Different Problems

<u>Fossil fuels</u> are burnt to release <u>energy</u>. We burn fossil fuels to power <u>vehicles</u> and to produce electricity in <u>power stations</u>. The <u>carbon-based</u> products of burning fossil fuels often <u>pollute</u> the <u>atmosphere</u>.

1) All fossil fuels contain large amounts of the element <u>carbon</u>, so it's no surprise that these fuels produce a lot of <u>pollutants</u> that contain <u>carbon</u>.

2) If the fuel is burnt where there's <u>lots of oxygen</u> available, then nearly all the carbon ends up in <u>carbon dioxide</u>. This adds to the carbon dioxide that's found naturally in the <u>atmosphere</u>.

3) If there's <u>not much oxygen</u> available, such as in a car <u>engine</u>, then small amounts of <u>carbon monoxide</u> and small particles of <u>carbon</u> are produced as well.

4) <u>Carbon dioxide</u>, <u>carbon monoxide</u> and <u>carbon particles</u> are all <u>pollutants</u>.

Carbon Dioxide

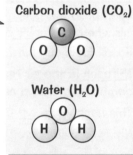

Carbon dioxide (CO_2)

Water (H_2O)

1) <u>Carbon dioxide</u> has the formula CO_2, which means that a <u>molecule</u> of carbon dioxide has <u>two oxygen</u> atoms and <u>one carbon</u> atom.

2) In the same way, <u>water's</u> formula, H_2O, tells us that a <u>molecule</u> of water is made of <u>two hydrogen</u> atoms and <u>one oxygen</u> atom.

3) Carbon dioxide (CO_2), like any other atmospheric pollutant, will <u>stay</u> in the <u>atmosphere</u> causing problems until it's <u>removed</u>.

4) CO_2 can be <u>removed</u> from the atmosphere naturally. Plants <u>use up</u> CO_2 from the air when they <u>photosynthesise</u>. CO_2 also <u>dissolves</u> in rainwater and in seas, lakes and rivers.

5) Despite these ways of <u>removing</u> CO_2 from the atmosphere, CO_2 levels can still <u>increase</u> if human activity, e.g. burning fuels, adds extra CO_2 into the atmosphere.

6) Increased CO_2 levels increase the <u>greenhouse effect</u>, which is <u>warming</u> up the Earth. This may change the world's <u>climate</u>, possibly causing <u>flooding</u> due to the <u>polar ice caps melting</u> (see p.59).

Carbon Monoxide

Unlike carbon dioxide, a <u>carbon monoxide</u> molecule has only <u>one oxygen</u> atom attached to a <u>carbon</u> atom. Carbon monoxide is produced if there's not enough <u>oxygen</u> available when fuels burn.

Carbon monoxide (CO)

Carbon monoxide is <u>poisonous</u> — if a dodgy boiler in your home starts giving out carbon monoxide, it can make you <u>drowsy</u> and <u>headachy</u>, and can even sometimes <u>kill</u>.

Particulate Carbon

Often tiny <u>particles</u> of <u>carbon</u> are produced when fuels burn — this is called <u>particulate carbon</u>. If they escape into the <u>atmosphere</u>, which they often do, they just <u>float</u> around. Eventually they fall back to the ground and deposit themselves as the horrible black dust we call <u>soot</u>.

A lot of soot just falls onto <u>buildings</u>, making them look <u>dirty</u>, like this one.

Problems problems... there's always summat goin' wrong...

<u>Energy companies</u> are aware of the problems of burning fuels so they're investing money into developing <u>cleaner</u>, <u>renewable</u> sources of energy, which might not save the world but it'll help a lot. Hooray!

Air Pollution — Sulfur and Nitrogen

So, now you know about the different kinds of carbon pollution. On to sulfur and nitrogen...

Sulfur Pollution Comes from Impurities in Fuels

1) Many of our fuels are hydrocarbon-based, like petrol and natural gas. Some are just carbon-based, like coal. These fuels contain loads of impurities, including sulfur.

2) When the fuel burns, the sulfur burns too — the sulfur atoms combine with the oxygen in the air to produce the pollutant sulfur dioxide.

> Sulfur dioxide has the formula SO_2. This tells you that a sulfur dioxide molecule is two oxygen atoms and one sulfur atom.

Sulfur dioxide (SO_2)

3) So when power stations and vehicle engines burn fossil fuels like coal and oil, small amounts of the pollutant sulfur dioxide are produced. This sulfur dioxide usually ends up in our atmosphere.

Nitrogen Pollution Involves Nitrogen from the Air

Nitrogen pollution doesn't actually come from the fuel itself — it's formed from nitrogen in the air when the fuel is burnt.

1) Fossil fuels can burn at such high temperatures that nearby atoms in the air react with each other.

2) Nitrogen in the air reacts with the oxygen in the air to produce small amounts of compounds known as nitrogen oxides — nitrogen monoxide and nitrogen dioxide.

3) This happens in car engines.

4) Nitrogen oxides are pollutants, and are usually spewed straight out into the atmosphere.

> 1) Nitrogen monoxide has the formula NO — it's made of one nitrogen and one oxygen atom.
>
> 2) Nitrogen dioxide has the formula NO_2 — it is made of two oxygen atoms joined to one nitrogen atom.

Nitrogen monoxide (NO)

Nitrogen dioxide (NO_2)

Sulfur and Nitrogen Pollution Causes Acid Rain

1) As with other pollutants, when sulfur dioxide and nitrogen oxides get into the atmosphere they will stay there until something gets rid of them.

2) The way sulfur dioxide and nitrogen oxides usually leave our atmosphere is in the form of acid rain.

3) When the sulfur dioxide emitted from vehicle engines and power stations reacts with the moisture in clouds, dilute sulfuric acid is formed.

4) In the same way, nitrogen oxides react with moisture in the atmosphere to produce dilute nitric acid.

5) Eventually, much of this acid will fall as acid rain, which is bad news for buildings, plants, animals and humans.

Acid rain makes statues talk too.

6) Acid rain causes lakes to become acidic, killing plants and animals. It also kills trees and damages buildings and statues made from some kinds of stone, e.g. limestone.

Acid rain — well at least it's more exciting than carbon...

Sulfur, for those of you who might think otherwise, is spelt with an 'f'. So don't go writing 'sulphur' any more because we're spelling it the international way these days. It wasn't my decision, because unfortunately that sort of thing isn't up to me. So, from here on, it's sulfur to us both.

Reducing Pollution

There are plenty of things we can be doing to try and reduce air pollution.

We Can Reduce Pollution from Power Stations...

1) The less <u>electricity</u> we use, the less <u>fossil fuel</u> will need to be burnt in power stations, and the less <u>pollution</u> will be created.

2) Much of the <u>sulfur</u> can be taken out of <u>natural gas</u> and the <u>fuel oil</u> that power stations use. This means that little <u>sulfur dioxide</u> will be produced when it burns.

3) When <u>coal</u> is burnt in power stations, most of the <u>sulfur dioxide</u> and the <u>particulates</u> (carbon particles and ash) can be removed before they can get into the atmosphere.

4) Don't forget — the only way to <u>reduce CO_2 emissions</u> is to reduce the amount of <u>fossil fuels</u> burnt.

...And from Our Cars

1) Motor vehicles now have more <u>efficient</u> engines, which burn <u>less fuel</u> and so create <u>less pollution</u>.

2) <u>Low-sulfur fuel</u> for cars is now available, which means <u>less sulfur dioxide</u> is emitted from the exhaust.

3) Many cars are now fitted with <u>catalytic converters</u>. These convert harmful <u>nitrogen monoxide</u> into harmless <u>nitrogen</u> and <u>oxygen</u>. They also convert the very toxic gas <u>carbon monoxide</u> into the less harmful gas <u>carbon dioxide</u>.

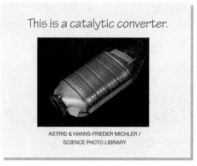

This is a catalytic converter.

ASTRID & HANNS-FRIEDER MICHLER / SCIENCE PHOTO LIBRARY

Cars more than 3 years old must pass an MOT test once a year to prove they're safe and not too polluting.

4) If everyone used <u>public transport</u> instead of individual cars then less <u>petrol</u> would be burnt overall.

5) There's a <u>legal limit</u> on the amount of polluting <u>emissions</u> that cars can give out. A car's emissions are checked in the <u>MOT</u> test.

Laws and Regulations Help Reduce Pollution

Certain <u>laws</u> and <u>regulations</u> exist to make individuals and organisations <u>responsible</u> for the <u>pollution</u> they produce.

> Any process that is likely to add <u>pollutants</u> to the <u>atmosphere</u>, including scientific research, must comply with laws and regulations that <u>limit</u> the amount of polluting chemicals that can be released into the atmosphere (and elsewhere).

My brother is a major source of air pollution...

So... it doesn't sound like reducing pollution is going to be easy. We need to reduce the <u>demand</u> for electricity by not <u>using</u> so much of it, and then we won't need so many <u>power stations</u>. Start small by remembering to switch off lights, and don't leave stuff on <u>stand-by</u> — that's just wasting electricity.

Interpreting Pollution Data

Here's a page on the reliability of scientific research and why you shouldn't believe everything you read in the papers.

Claims About Air Pollution Can be Hard to Justify

Lots of studies have linked air pollution to various health problems.
When these claims are made though, we shouldn't just take the results at face value.

> Say some scientists have claimed that the number of asthma attacks in children is linked to the level of a particular pollutant in the air.

Don't be too convinced about the results if...

...It's Just One Small Study That Hasn't been Repeated

1) The scientists would have to provide details of the experiments they did, the size of their sample and the results they got. The bigger the sample used in the study, the more confident you can be about the results.

2) That's not enough though. Other scientists would need to repeat the experiments and get the same results before the claims could be accepted by the scientific community.

3) A studys' results are only really trustworthy if other scientists can replicate them.

...It Doesn't Take Other Factors into Account

Any outcome (e.g. an asthma attack) could be caused by factors other than the one you're investigating (e.g. air pollution) — air pollution isn't the only cause of asthma attacks.

1) It's thought that asthma attacks can be triggered by factors like high pollen levels, infections (e.g. colds and flu) and emotional stress.

2) Any study into the effect of air pollution levels on the number of asthma attacks needs to account for these factors in some way, e.g.

Pollen — it has a lot to answer for.

- The study should be carried out at a time of year when the pollen levels are approximately the same in high pollution and low pollution areas — so probably not spring or summer.

- If any of the subjects get over-stressed or ill in a way that might affect the results, they should be excluded from the study.

...The Conclusion Goes Beyond the Data

1) A correlation between two things doesn't necessarily mean that one caused the other. So a correlation between air pollution levels and the number of asthma attacks doesn't necessarily mean that the air pollution is causing the attacks.

2) Also remember that if one thing does cause something else, it doesn't mean that it will definitely make it happen. So if the study did show that air pollution causes asthma attacks, all that means is that your risk of having an asthma attack is increased in a high pollution area.

There's a correlation between madness and chemistry knowledge...

Don't just believe what you read — ask questions like 'but how reliable was the experiment?', 'but could anything else have caused those results?' and 'but does that correlation show causation?'. Have an enquiring mind. This is how science works — scientists look for evidence for or against theories.

Sustainable Development

Sustainable development is a hot topic at the moment — it's relevant to subjects like Geography too.

More People Means More Pressure on the Environment

1) The world's population is increasing at an amazing rate.
2) It has doubled in the last 40 years and now totals 6.5 billion. It shows no sign of slowing down.
3) The more of us there are, the more food must be produced, the more fuel is burnt and the more land is used up. This all puts a lot of pressure on the environment.

Scientific Advances May Have Helped Cause the Problems...

Science-based technology has improved the quality of life for most of us, but is also unfortunately contributing to environmental problems. Computers are a good example of this:

1) Computers have made workplaces more efficient, and the internet has improved global communication, which is great for both individuals and businesses. Computers are still getting better all the time.
2) The trouble is that computers are full of poisonous substances. Computers don't last for ever and these poisonous substances cause a problem when it comes to computer disposal.
3) Many old computers are sent to poorer countries to be dismantled. There they're often dumped in landfills. This poisons the land for the people, animals and plants that live there.

> Computers are great, as are many of the scientific advances that make our lives easier.
> However, the cost to the environment and ultimately to humans is massively important.
> Benefits need to be weighed against costs when deciding whether to use new technologies.

...But Other Advances Could Help Us Live More Sustainably

Often science comes up with solutions to the problems it causes. In particular, science has helped us find ways of using our natural resources to meet our needs without messing things up so that future generations can't meet theirs. This is known as sustainable development. For example:

1) Scientific advances have allowed us to develop buildings that use less energy. For example, modern buildings have insulated windows and doors which mean less energy is used to heat them.
2) Using less energy means less fossil fuels are burnt in the building or at power stations and so less pollution reaches the atmosphere.
3) Buildings can also use renewable resources as energy sources, e.g. solar energy can be used to heat water and generate electricity. Wind turbines on the roofs of houses can be used to generate electricity.

> These days, there are laws and regulations governing scientific research.
> In many areas of science, scientists have to show that what they're
> developing will not go on to have an unacceptable impact on the environment.
> This particularly applies to pollutants being released into the atmosphere.

Try wind power — it can blow your mind...

It's kind of obvious that scientific development has led to environmental problems. I mean, back when there was little industry but lots of farming, fishing and stuff, there was little pollution. But since then, lots of scientific advances have caused problems — from pollution to cutting down rainforests.

Revision Summary for Module C1

Congratulations! You've made it through C1. You should now be an expert on fossil fuels, carbon, sulfur, air pollution and the general problems we're causing to the environment and our own health. You should also have picked up a little bit on atoms and what happens to them in chemical reactions. Anyway, maybe it's time you found out how much you do know. A good try at the questions below should give you some idea. Here we go...

1) What three main gases is the atmosphere made of and in what proportions are they found?

2) What is a hydrocarbon?

3) Name the main elements that make up coal.

4) When a hydrocarbon fuel burns, with what substance in the air do the hydrogen and carbon atoms combine?

5) When hydrocarbons burn in plenty of oxygen, the atoms involved rearrange themselves into carbon dioxide and what else?

6) Atoms can disappear completely in some chemical reactions — true or false?

7) Are the properties of reactants and products always the same or can they be different?

8) Name one result of pollution which causes direct harm to humans, and one which causes indirect harm.

9) What removes carbon dioxide from the atmosphere?

10) What atoms make up carbon monoxide? Under what conditions is carbon monoxide produced?

11) What are particles of carbon otherwise known as and what kind of pollution do they cause?

12) Describe briefly how the pollutant sulfur dioxide is produced.

13) How does sulfur dioxide leave the atmosphere?

14) What effects does acid rain have on the environment?

15) How is nitrogen dioxide produced?

16) Give the formulas of nitrogen monoxide and nitrogen dioxide.

17) What effect do nitrogen oxides have on the environment?

18) What do catalytic converters do?

19) Name two things that everyone could do in order to reduce carbon dioxide pollution.

20) How do MOT tests help combat air pollution?

21) If both air pollution and cancer increase, does that mean air pollution causes cancer?

22) Why is it a problem that the human population is rapidly increasing?

23) Give an example of a scientific advancement that contributes to environmental problems.

24) What is sustainable development?

25) Give an example of a renewable resource that could be used to provide energy for our homes.

The Changing Earth

There are some really <u>dull</u> things in the Solar System. For example, <u>asteroids</u> don't do much except go round and round the Sun. The <u>Earth</u>, however, is <u>different</u>...

The Earth is an Active Planet

1) It's kind of tempting to think that the Earth is a <u>steady</u>, <u>unchanging</u> place that'll always look pretty much as it does now.

2) But no... in a few tens of millions of years, Earth's going to look a <u>lot</u> different.

3) <u>Mountains</u> that are <u>enormous</u> today won't be nearly so grand. And the map of the world will be very different — whole <u>continents</u> will have <u>moved</u>. Weird.

4) This is nothing new — Earth's been changing for <u>thousands of millions</u> of years.

Rocks Show Changes in the Earth

Rocks <u>change</u> over the years — so the surface of the Earth hasn't always looked the way it does now...

1) Look at the <u>Grand Canyon</u>, for example. Over time, the Colorado River has <u>eroded</u> (worn away) the rock, leaving an impressive slash through the middle of <u>Arizona</u>.

2) Erosion goes on <u>everywhere</u>. We see it happening, for example, when <u>cliffs</u> are worn away by the <u>sea</u>. But <u>other</u> processes must be happening as well. If not, all the high ground would have been <u>worn down</u> by now... Earth would be <u>perfectly smooth</u>.

DR MORLEY READ / SCIENCE PHOTO LIBRARY

3) So some process must be making <u>new</u> mountains. And <u>evidence</u> is pretty easy to find — e.g. when <u>lava</u> from <u>volcanoes</u> sets, it forms brand new rock.

4) There are also some pretty spectacular <u>forces</u> at work in the Earth, which can push up old rock to make new mountains — some rock formations show rock that's been squeezed so hard it's just <u>folded</u>.

5) More evidence that rock is constantly forming comes from <u>fossils</u>. The animals and plants couldn't have <u>dug themselves</u> into the <u>middle</u> of rocks — the <u>rocks</u> must have <u>formed</u> around them. This type of rock is made by <u>sedimentation</u> — particles eroded from existing rock get washed to the sea and settle as sediment. Over time, these sediments get crushed together to make <u>new rock</u>.

Fossils are traces of animals and plants from long ago, usually found in rocks.

All the rock processes that we can see happening <u>now</u> — e.g. <u>erosion</u>, <u>sedimentation</u>, <u>volcanic</u> activity, etc. will have been going on for <u>millions</u> of years. Geologists can use that fact to explain how various <u>modern landscapes</u> were formed.

Scientists Can Use Radioactivity to Tell How Old Rocks Are

1) By looking at the proportions of <u>radioactive</u> potassium-40, and the element it <u>decays</u> into, in certain rocks, scientists can get a good idea of when the rocks were <u>formed</u>. (It's tricky, but it can be done.)

2) Scientists can use this to help estimate the age of the planet — the Earth must have been around for <u>at least</u> as long as the <u>oldest</u> rocks.

If you want to see the Rocky Mountains, don't leave it too long...

<u>Erosion</u> will eventually wear down the <u>Rockies</u> — in a few million years they <u>won't be there</u>. Shame. There's more evidence about <u>how</u> and <u>why</u> Earth changes over the page. It's all exciting stuff.

Observations and Explanations

Observations and explanations aren't the same. Anyone can observe something, but explaining it is trickier.

Observations About the Earth Hadn't been Explained

1) For years, fossils of very similar plants and animals had been found on opposite sides of the Atlantic Ocean.

2) Other things about the Earth also puzzled people — like why the coastlines of Africa and South America matched so well. And why fossils from sea creatures had been found high in the Alps.

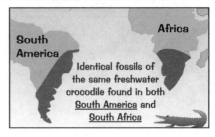

South America | Africa

Identical fossils of the same freshwater crocodile found in both South America and South Africa

Explaining These Observations Needed a Leap of Imagination

What was needed was a scientist with a bit of insight... a smidgeon of creativity... a touch of genius...

1) Alfred Wegener hypothesised that Africa and South America had previously been one continent which had then split. He started to look for more evidence to back up his hypothesis...

2) E.g. he found that there were matching layers in the rocks on different continents, and similar earthworms living in both South America and South Africa.

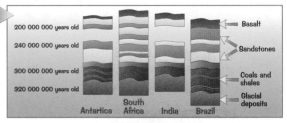

200 000 000 years old
240 000 000 years old
300 000 000 years old
320 000 000 years old

Antartica | South Africa | India | Brazil

Basalt
Sandstones
Coals and shales
Glacial deposits

N. America
Africa
S.America

3) Wegener's theory of 'continental drift' supposed that there had once been a single 'supercontinent' — which he called Pangaea. According to Wegener, Pangaea broke into smaller chunks... and these chunks (our modern-day continents) are still slowly 'drifting' apart.

The Theory Wasn't Accepted at First — for a Variety of Reasons

1) Wegener's theory explained things that couldn't be explained by other, competing theories (e.g. the formation of mountains — which Wegener said happened as continents smashed into each other). But the reaction from other scientists was generally hostile.

2) The main problem was that Wegener's explanation of how the 'drifting' happened wasn't convincing (and the movement wasn't detectable). Wegener claimed the continents' movement could be caused by tidal forces and the Earth's rotation — but other geologists showed that this was impossible.

3) Also, it probably didn't help that Wegener wasn't a 'proper' geologist — he was a meteorologist.

Eventually, the Evidence Became Overwhelming

1) In the 1950s, scientists investigated the Mid-Atlantic ridge.

2) They found that magma (molten rock) rises up through the sea floor, solidifies and forms underwater mountains... new sea floor was being created.

3) This was convincing evidence that the continents were moving apart.

Symmetrical undersea mountains made of basalt
Ocean Floor
Magma
Oceanic Plates moving apart

I told you so — but no one ever believes me...

Wegener wasn't right about everything, but his main idea was correct. Nowadays, we know that it's not just the continents that move, but whole tectonic plates (including ocean floors) — see the next page.

Module P1 — The Earth in the Universe

The Structure of the Earth

We can tell a lot about what goes on <u>deep inside</u> the Earth by looking at what happens on the <u>surface</u>.

The Earth Has a Crust, Mantle and Core

The Earth is <u>almost spherical</u> and it has a <u>layered</u> structure, a bit like a scotch egg. Or a peach.

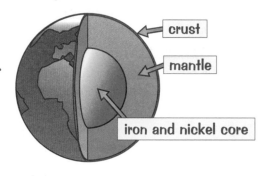

1) The bit we live on, the <u>crust</u>, is very <u>thin</u> (averaging about 20 km). There are <u>two types</u> of crust — <u>continental crust</u> (forming the land) and <u>oceanic crust</u> (under oceans).

2) Below that is the <u>mantle</u>. The <u>mantle</u> has all the properties of a <u>solid</u>, except that it can flow very <u>slowly</u>.

3) Within the mantle, <u>radioactive decay</u> takes place. This produces a lot of <u>heat</u>, which causes the mantle to <u>flow</u> in <u>convection currents</u>.

4) At the centre of the Earth is the <u>core</u>, which we think is mainly made of <u>iron and nickel</u>.

The Earth's Surface is Made Up of Tectonic Plates

1) The crust and the upper part of the mantle are cracked into a number of large pieces called <u>tectonic plates</u>. These plates are a bit like <u>big rafts</u> that 'float' on the mantle.

2) The plates don't stay in one place though.

3) The map shows the <u>edges</u> of the plates as they are now, and the <u>directions</u> they're moving in (red arrows).

4) Most of the plates are moving at speeds of <u>a few cm per year</u> relative to each other.

5) Occasionally, the plates move very <u>suddenly</u>, causing an <u>earthquake</u>. <u>Volcanoes</u> also often occur at the boundaries between two tectonic plates.

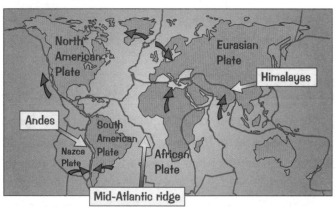

Scientists Try to Predict Earthquakes and Volcanic Eruptions

1) Tectonic plates can stay put for a while and then <u>suddenly</u> lurch forwards without warning to cause an <u>earthquake</u>. <u>Volcanoes</u> can be very <u>unpredictable</u> too.

2) Scientists are trying to find <u>clues</u> that an earthquake or volcanic eruption might happen soon — things like <u>strain</u> in underground rocks or <u>magma movement</u> under a volcano.

3) But it's tricky. Most likely, scientists will only be able to say that an earthquake or volcanic eruption's <u>more likely</u> — but not that it's <u>certain</u>.

4) However, knowing even just a <u>little</u> about whether an earthquake or eruption is likely could <u>save lives</u>. If disaster's likely to strike <u>soon</u> (or even <u>soonish</u>), the authorities can <u>warn residents</u> and <u>evacuate</u> the area.

Learn about plate tectonics — but don't get carried away...

It's important to remember that earthquakes and volcanoes are <u>unpredictable</u>, even with the <u>best</u> equipment in the world. However, even a <u>little</u> information about the likelihood of an eruption or earthquake would be potentially useful. Learn <u>all</u> the information — you're going to need it.

The Solar System

What a fantastic place Earth is. But the rest of our Solar System is pretty cool too.

Planets Reflect Sunlight and Orbit the Sun in Ellipses

Our Solar System consists of a star (the Sun) and lots of stuff orbiting it in slightly elongated circles.

- Closest to the Sun are the inner planets — Mercury, Venus, Earth and Mars.
- Then the asteroid belt — see below.
- Then the outer planets, much further away — Jupiter, Saturn, Uranus, Neptune (and Pluto).
- There are also various other things — comets, dust, meteors and so on... all in orbit around the Sun. These also count as part of our Solar System.

Stars and Planets are Very Different from Each Other

1) You can see some planets with the naked eye. They look like stars, but they're totally different.
2) Stars are huge (the Sun's diameter is over 100 times bigger than the Earth's), very hot and very far away. They give out lots of light — which is why you can see them even though they're far away.
3) Planets are smaller and they just reflect sunlight falling on them. The planets in the Solar System are also much much closer to us than any star (except the Sun).
4) Planets often have moons orbiting them. Jupiter has at least 63 of 'em. We've just got one.

The Solar System Formed from a Big Gas and Dust Cloud

1) The Solar System didn't appear overnight — it formed over a long time from a cloud of gas and dust.
2) For some reason (maybe a nearby star exploding), this dust cloud started to get squeezed slightly.
3) Once the particles had moved a bit closer to each other, gravity took over. It pulled things closer and closer together until the whole cloud started to collapse in on itself.
4) At the centre of the collapse, particles came together to form a protostar, which went on to become our Sun (see page 32).
5) Elsewhere, dust started to clump together — and these clumps became planets.
6) So the Sun and the planets formed at about the same time — they have similar ages.

Asteroids and Comets are Smaller Than Most Planets

1) Asteroids and comets are made of stuff left over from the formation of the Solar System.
2) The rocks between Mars and Jupiter didn't form a planet, but stayed as smallish lumps of rubble and rock — these are asteroids.

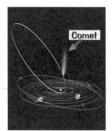

3) Comets are balls of rock, dust and ice which orbit the Sun in very elongated ellipses, often in different planes from the planets. The Sun is near one end of the orbit.
4) As a comet approaches the Sun, its ice melts, leaving a bright tail of gas and debris which can be millions of kilometres long. This is what we see from Earth.

Asteroids... my dad had those — very nasty...

So one minute there was a big cloud of gas and dust... the next there were planets, a star, asteroids, comets and all sorts. Okay... it actually took hundreds of millions of years, but it's still pretty impressive.

Danger from Space

If anything from space hit the Earth, it could be unpleasant. The atmosphere provides some protection against smaller things, but if something bigger hit, the consequences could be bad.

Earth is Sometimes Hit by Rocks and Dust

1) There's a lot of stuff floating about in space — ranging in size from specks of dust to huge asteroids.

2) Quite often, small particles of dust or small rocks enter the Earth's atmosphere.
   On their way through the atmosphere they usually burn up — and we see them as 'shooting stars'.

3) Sometimes, not all of the object burns up and part of it crashes into the Earth's surface.
   This only happens rarely, but when large things do hit us, they can cause havoc...

   - They can cause huge tsunamis if they land in the ocean. On land they can start fires, and throw loads of hot rocks and dust into the air.
   - The dust and smoke from a large impact can block out the sunlight for many months, causing climate change — which in turn can cause species to become extinct.
   - They also make big holes in the ground (craters, if we're being technical).

   ©iStockphoto.com/Stephan Hoerold

   > The picture shows an impact crater in Arizona. The theory that this was the result of a collision with an object from space took a while to be accepted (as is often the case with new claims). But evidence eventually emerged to convince the doubters.

4) We can tell that asteroids have collided with Earth in the past. There are the big craters, but also:
   - layers of unusual elements in rocks — these must have been 'imported' by an asteroid,
   - sudden changes in fossil numbers between adjacent layers of rock, as species suffer extinction.

Impact was about here...

> About 65 million years ago an asteroid about 10 km across struck the Yucatán peninsula in Mexico. The dust it kicked up caused global temperatures to plummet, and over half the species on Earth subsequently died out (including maybe the last of the dinosaurs).

A Really Big Impact is Very Unlikely in Your Lifetime

1) Astronomers use powerful telescopes to search for and monitor asteroids or comets which might be on a collision course with Earth. Then they can calculate an object's likely trajectory (the path it's probably going to take) and find out if it's actually anything to worry about.

2) Scientists can also get an idea about the likelihood of something hitting Earth by looking at the Moon. The Moon has no atmosphere for objects to burn up in, and so impact craters are much more common on the Moon than on Earth. Also, the Moon's craters don't get eroded by water or wind, or disturbed in other ways, e.g. by volcanoes.

3) In the last 600 million years, Earth has probably been hit by 60 or so objects more than 5 km across. And an impact like the one in the Yucatán peninsula only happens about once every 300 million years.

4) So the odds of a really big impact with Earth within the next century are tiny.

5) Smaller impacts are more frequent. An object of the size that made the Arizona crater hits every 160 years or so. Something that size wouldn't cause worldwide devastation, though (but it would trash the immediate surroundings).

Don't go to the Moon for a night out — there's no atmosphere...

When a big object's discovered that might be heading in Earth's direction, the newspaper headlines tend to be fairly dramatic. But you need to keep this kind of thing in some kind of perspective... For example, an object was found a few years ago that had a 1 in 40 chance of hitting Earth. But remember, the odds were 39 in 40 (i.e. 97.5%) that it wouldn't hit, which is far more likely (though less newsworthy).

Beyond the Solar System

There's all sorts of exciting stuff out there. The whole Solar System is just part of one galaxy.
And there are billions upon billions of galaxies. Yup, the Universe is big — huge in fact...

We're in the Milky Way Galaxy

1) Our Sun is one of many billions of stars which form the Milky Way galaxy.
The Sun is about halfway along one of the spiral arms of the Milky Way.

2) The distance between neighbouring stars in a galaxy is usually millions
of times greater than the distance between planets in the Solar System.

3) And the diameter of the Milky Way is about 600 billion times the
diameter of the Sun. Yup... it's pretty big.

The Whole Universe Has More Than a Billion Galaxies

1) Galaxies themselves are often millions of times further apart
than the stars are within a galaxy.

2) So even the slowest among you will have worked out that the
Universe is mostly empty space and is really really BIG.

3) The Universe is also really really old — about 14 billion years
old, according to scientists' latest theories (see p.33).

Distances in Space Can be Measured Using Light Years

1) Once you get outside our Solar System, the distances between stars and between galaxies are
so enormous that kilometres seem too pathetically small for measuring them.

2) For example, the closest star to us (after the Sun) is about 40 000 000 000 000 kilometres away
(give or take a few hundred billion kilometres). Numbers like that soon get out of hand.

3) So we use light years instead. A light year is the distance that light travels through a vacuum
(like space) in one year. Simple as that. 1 light year is equal to about 9 460 000 000 000 km.

4) Just remember — a light year is a measure of DISTANCE (not time).

There Might be Other Life in the Universe

1) It's possible that life exists elsewhere in the Universe. After all, when you consider how huge the
Universe is, why should our planet be the only place with suitable conditions...

2) Scientists believe conditions necessary for life (e.g. water, nutrients...) are most likely to be found
on other planets, or on moons. So they started a search for planets outside the Solar System.

3) So far, about 185 planets have been discovered orbiting nearby stars. It's very likely there are a lot
more, given how many stars there are. Maybe some of these planets or their moons could be
inhabited. But so far it's just a theory — we haven't found other life yet.

4) Some scientists believe the Sun and the Earth are actually pretty special, e.g. the Sun's position in
the Milky Way keeps us away from too much radiation, and the Earth's unusual tectonic activity is
a great way to recycle the carbon needed for life. These features are very helpful but not
necessarily that likely, and so suitable conditions for life may not be as common as you'd think.

Spiral arms — would you still need elbows...

Until we actually discover life elsewhere, the debate about whether it exists or not is likely to continue.
Proving life does exist is relatively 'easy' — you find life and the argument's over. But it's impossible to
prove beyond doubt that life doesn't exist elsewhere — after all, there are plenty of places it could be.

Looking into Space

We can't <u>travel to</u> stars to study them — it'd take 'a while' (thousands of years, at the very least). All we can realistically do is measure the <u>radiation</u> coming <u>from</u> them.

Radiation Can Tell Us a Lot About Stars and Galaxies

1) We can tell a lot about a star by studying the <u>electromagnetic radiation</u> (e.g. light, X-rays, radio waves — see p.54) it emits. For example, the <u>colour</u> that a star appears is actually a pretty good guide to its <u>temperature</u>.

Big dish telescopes like these are for detecting radio waves.

2) To work out <u>how far away</u> a star is, you can use various methods.

3) For 'nearby' stars, you can use <u>parallax</u>. Astronomers take <u>pictures</u> of the sky 6 months apart (when Earth is at <u>opposite sides</u> of its orbit).

<u>Parallax</u> is the <u>apparent movement</u> of something when you look at it from <u>different places</u> (e.g. hold your finger at arm's length and look at it first through your left eye, then your right — it seems to move against the background).

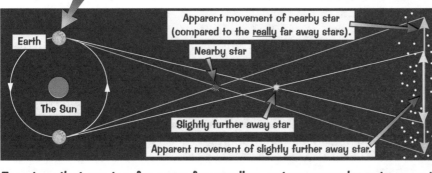

Earth

Nearby star

Apparent movement of nearby star (compared to the <u>really</u> far away stars).

The Sun

Slightly further away star

Apparent movement of slightly further away star.

The <u>apparent movement</u> of a star between the two photos lets you work out <u>how far away</u> it is. Stars <u>further away</u> appear to move less (the <u>really</u> distant stars don't appear to move at all — the movement is too small to detect).

4) For stars that are too far away for parallax, astronomers have to use other, <u>less trustworthy</u> methods.

The Atmosphere and Light Pollution Cause Some Problems

1) If you're trying to detect <u>light</u>, Earth's <u>atmosphere</u> can be a bit of a pain — it <u>absorbs</u> quite a bit of the light coming from space <u>before</u> it can reach us.

2) And <u>light pollution</u> (light thrown upwards from streetlamps, etc.) makes it <u>hard</u> to see <u>dim</u> objects.

3) That's why scientists put the <u>Hubble Space Telescope</u> in <u>space</u> — where you don't get these problems.

We See Stars and Galaxies as They Were in the Past

1) <u>Light</u> travels pretty <u>fast</u> — but it <u>does</u> take time to get from one place to another.

2) Radiation from the Sun takes about <u>8 minutes</u> to <u>reach us</u>.

3) That means that when we look at the Sun, we see it as it was <u>about 8 minutes ago</u>. So if it suddenly exploded (but, fingers crossed, it won't for a while), we wouldn't know <u>anything</u> about it for <u>about 8 minutes</u>.

4) Since the nearest star to us after the Sun is about <u>4.2 light years</u> away, light from it takes <u>4.2 years</u> to reach us. This means we see it as it was <u>4.2 years ago</u>.

5) When we look at other stars, this effect is even more extreme. For example, we see the <u>North Star</u> as it was during the time of <u>William Shakespeare</u>.

Constant stars, in them I read such art... (From Shakespeare's Julius Caesar.)

A bit of culture there. Now then... measuring the distance to a star is <u>very hard</u> (especially ones that are really far away) — scientists have to make certain <u>assumptions</u> about the <u>star</u> and about the <u>space</u> between it and Earth. What this means is that there's a <u>degree of uncertainty</u> in those measurements. So if someone says they know to the <u>nearest kilometre</u> how far away a star in a distant galaxy is, they're lying.

The Life Cycle of Stars

Stars go through <u>many traumatic stages</u> in their lives — you don't need to learn the details, but you do need to know that <u>all stars</u> go through a life cycle something like this...

Clouds of Dust and Gas

1) Stars <u>initially form</u> from <u>clouds of DUST AND GAS</u>.

Protostar

2) <u>Gravity</u> makes the gas and dust <u>spiral together</u> to form a hot ball called a <u>protostar</u>.

Stable
Star

3) When the <u>temperature</u> gets <u>high enough</u>, a process called <u>fusion</u> starts — hydrogen atoms join together to make helium. This gives out massive amounts of <u>heat and light</u>... a star is born. It immediately enters a <u>long stable period</u> that can last <u>several billion years</u>. (The Sun is in the middle of this stable period — or to put it another way, the <u>Earth</u> has already had <u>half its innings</u> before the Sun <u>engulfs</u> it!)

Red Giant

4) Eventually the star runs out of hydrogen and <u>swells</u> into a <u>RED GIANT</u>.

Small stars

White Dwarf

Big stars

5) A small-to-medium-sized star like the Sun then <u>throws off</u> its <u>outer layers</u> to leave behind a hot, dense core — a <u>WHITE DWARF</u>, which cools down and eventually disappears (awww...).

Neutron
Star...

...or
Black
Hole

Supernova

6) <u>Big stars</u>, however, eventually <u>explode</u> in a <u>SUPERNOVA</u>.

7) The <u>exploding supernova</u> leaves behind a <u>very dense core</u> called a <u>NEUTRON STAR</u>. If the star is <u>big enough</u> this will become a <u>BLACK HOLE</u>.

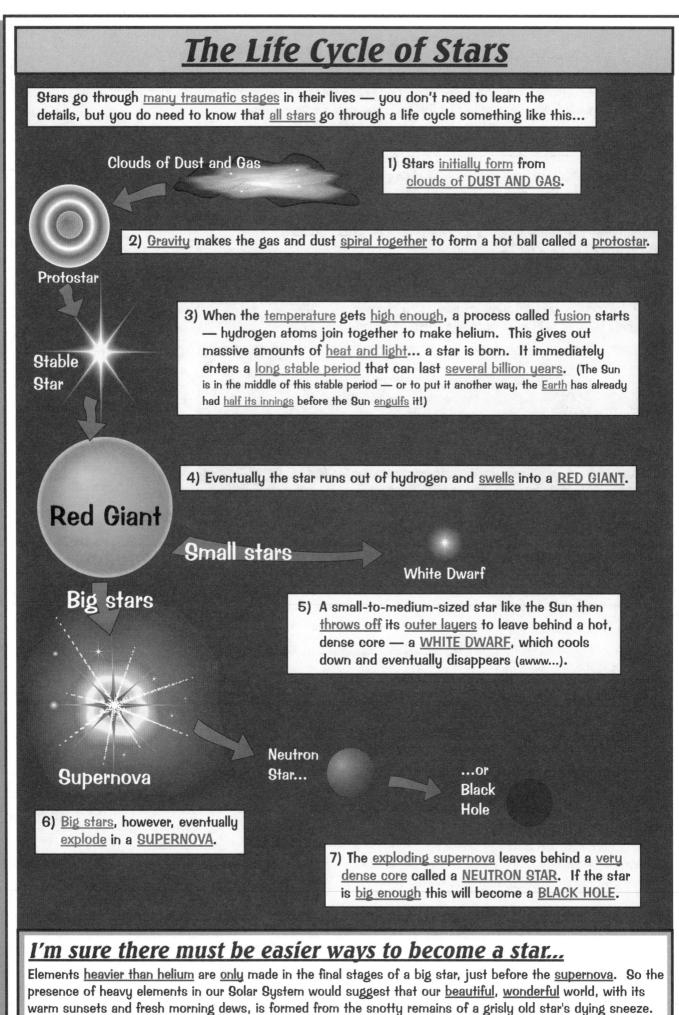

I'm sure there must be easier ways to become a star...

Elements <u>heavier than helium</u> are <u>only</u> made in the final stages of a big star, just before the <u>supernova</u>. So the presence of heavy elements in our Solar System would suggest that our <u>beautiful</u>, <u>wonderful</u> world, with its warm sunsets and fresh morning dews, is formed from the snotty remains of a grisly old star's dying sneeze.

The Life of the Universe

They expect you to know all about the history of the Earth, the Solar System, stars... and the <u>Universe</u>. Sheesh... they don't give these GCSEs away easily, do they.

The Universe Seems to be Expanding

1) The Universe is pretty big already. But it looks like it's getting <u>even bigger</u> all the time.
2) All its <u>galaxies</u> seem to be moving away from each other.
3) The <u>conclusion</u> appears to be that the whole Universe is <u>expanding</u>.

The Evidence Suggests the Universe Started with a Bang

All the galaxies are moving away from each other at great speed... suggesting something must have <u>got them going</u> in the first place. That 'something' was probably a <u>big explosion</u> — the <u>Big Bang</u>.

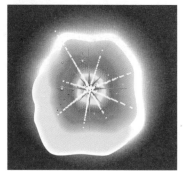

1) The Big Bang theory says that all the matter in the Universe initially occupied <u>a very small space</u> (<u>all</u> the matter in <u>all</u> the galaxies squashed into a space <u>much much smaller</u> than a pinhead — <u>wowzers</u>). Then it '<u>exploded</u>' — the space started expanding, and the <u>expansion</u> is still going on.

2) Using the Big Bang theory, we can estimate the <u>age</u> of the Universe. From the current <u>rate of expansion</u>, we think the Universe is about <u>14 billion years</u> old.

3) But estimating the age of the Universe is <u>very difficult</u> because it's hard to tell how much the expansion has <u>slowed down</u> since the Big Bang.

We Don't Know How (or If) the Universe Will End...

1) The Universe's ultimate fate depends on <u>how fast</u> it's expanding and the <u>total mass</u> there is in it.
2) We can measure the <u>expansion</u> quite easily (relatively speaking). But finding out <u>how much mass</u> there is in the Universe is a bit trickier.
3) Most of the mass appears to be <u>invisible</u> — it <u>doesn't glow</u> like a star. Astronomers can only detect this <u>dark matter</u> by the way it <u>affects the movement</u> of the things we <u>can</u> see.
4) The <u>amount</u> of dark matter in the Universe (as well as what it actually <u>is</u>) is one of the great unanswered questions in science. And it matters (no pun intended), because it's the amount of mass in the Universe that will dictate what happens to the Universe in the future.
5) This is because all the mass <u>everywhere</u> is attracted together by gravity. The more mass there is, the greater this pull, and the greater the slowing down of the Universe's <u>expansion</u>.

- If there's <u>enough mass</u> compared to <u>how fast</u> the galaxies are currently moving, the Universe will eventually <u>stop expanding</u> — and then <u>begin contracting</u>. This would end in a <u>Big Crunch</u>.
- If there's <u>not enough mass</u> in the Universe to stop the expansion, it could <u>expand forever</u>, with the Universe becoming <u>more and more spread out</u> into eternity.

In the beginning, there was... well, it's so hard to tell...

In fact (and this is a bit weird, I admit), according to recent observations, the Universe seems to be expanding <u>faster and faster</u>, <u>not</u> slowing down at all. What's going on there? We have no idea. But it shows how new evidence (if <u>confirmed</u> by other observations) makes scientists rethink their theories.

The Scientific Community

Who decides which scientific theories are sound, and which are nonsense... The 'scientific community', that's who. Let me explain, using the very simple (hmmm...) example of the beginning of the Universe...

Observations About the Universe Needed an Explanation

1) For years, many scientists (including Einstein) had believed that the Universe was static and unchanging (kind of how people thought of the Earth before Wegener).

2) So the observations that distant galaxies appear to be moving away from us were pretty important. This meant the Universe couldn't be unchanging — so a new theory was needed.

Different Scientists Came Up with Different Explanations

1) Scientists accepted that the Universe was expanding (similar results had been found before), and that an explanation was needed — but different people came up with different explanations.

2) One theory was the Big Bang theory — the Universe was originally squashed into a tiny space, which then exploded... and the expansion is still going on.

3) Another theory was the Steady State theory.
You don't need to learn the details, but basically, this said that the Universe had always existed (and would always exist) pretty much as it is now — there was no huge explosion. It agreed that the Universe was expanding, but said that it always had been. However, matter was being created in the gaps, so it never actually looked any different.

Scientists were Divided — Until New Evidence was Found

1) Now then... the scientific process depends on 'peer review'. This means that scientific theories are judged by the scientific community (all the world's scientists). If you can convince the scientific community, then your theory will be accepted (for the time being at least).

2) Traditionally, new findings and theories are announced in peer-reviewed journals, or at scientific conferences.

> A peer-reviewed journal is one where other scientists check results and theories before they're published. They check that people have been 'scientific' about what they're saying — e.g. that experiments and claims aren't biased. But this doesn't mean that the findings are correct, just that they're not wrong in an obvious kind of way.

3) Anyway... in 1959, opinion in the scientific community on the nature of the Universe was divided.
In a poll of astronomers, 11 were for the Big Bang, 8 for the Steady State, and 14 weren't sure what to believe. The way the scientific community jumped would depend on what other evidence came to light.

4) To test the Big Bang theory, scientists predicted that there should be some 'leftover radiation' from the initial explosion, and that it should be at a particular temperature. When they found it some years later, it was strong evidence that the Big Bang was the more likely explanation of the two.

5) Today nearly all astronomers agree there was a Big Bang. However, there are some who still believe in the Steady State theory.

Scientists Will Stick with the Big Bang Theory... *(Unless there's a better idea)*

1) The Big Bang theory isn't perfect. As it stands, it's not the whole explanation of the Universe — there are observations that the theory can't yet explain. But it's most likely the Big Bang theory will be adapted in some way to account for these rather than just dumped — it explains so much so well that scientists will need a lot of persuading to drop it altogether.

2) However, if someone comes up with a revolutionary new idea that explains the observations in a much more convincing way, then who knows...

Time and space — it's funny old stuff, isn't it...

Proving a scientific theory is impossible. If enough evidence points a certain way, then a theory can look pretty convincing. But that doesn't prove it's a fact — new evidence may change people's minds.

Revision Summary for Module P1

The only way you can tell if you've learned this module is to test yourself. Try these questions, and if there's something you don't know, go back and learn it. Even if it is all that tricky business about the origins of the Universe. And don't miss any questions out — you don't get a choice about what comes up on the exam so you need to be sure that you've learnt it all.

1) Describe one piece of evidence that shows that mountains are constantly being:
   a) worn down, b) made.

2) How can scientists tell how old some rocks are?
   What does this tell us about when the Earth was formed?

3) Describe two observations about the Earth that weren't properly explained before Alfred Wegener's theory.

4) How did Wegener's theory account for these observations?

5) Suggest why the initial reaction to Wegener's theory was hostile.

6) What evidence led scientists to accept most of Wegener's theory about the movement of continents?

7) Draw a labelled diagram showing the layered structure of the Earth.

8) What are tectonic plates? What can their movements cause?

9)* Why is it tricky to predict earthquakes accurately?

10) What's the difference between a planet and a star? And what are asteroids and comets?

11) How do scientists think the Solar System was formed?

12) How can we tell that objects from space have hit the Earth?
    How do scientists work out if another impact is likely soon?

13) What is the Milky Way? How big is it compared with the Sun?

14) What's a light year?

15) What can scientists currently say about life elsewhere in the Universe? Explain your answer.

16) Describe how scientists can tell how far away from Earth a nearby star is.

17) Why do we see stars as they were in the past?

18) What happens inside a star to make it so hot?

19) Describe the steps that lead to the formation of a main sequence star (like our Sun).

20) Describe the 'Big Bang' theory of the origin of the Universe.

21)* Who judges whether to accept or reject a new scientific theory? What is 'peer-review'?

* Answers on page 100.

Module P1 — The Earth in the Universe

Microorganisms and Disease

If you have <u>hypochondriac tendencies</u> then you're going to <u>love</u> this section — it's all about <u>diseases</u>, mainly infectious diseases and heart disease, and the different ways to prevent them.

Microorganisms Cause Many Diseases

1) Lots of <u>microorganisms</u> cause <u>disease</u>. Microorganisms that cause disease are called <u>pathogens</u>.

2) They include...

     Bacteria

     Protozoa (single-celled creatures)

     Fungi

     Viruses

3) Most microorganisms <u>reproduce</u> fastest in <u>warm</u>, <u>damp</u> places. This means that they tend to reproduce very quickly inside host organisms.

Marcus's fungal infection was way out of hand.

Symptoms Can Be Caused by Cell Damage or by Toxins

1) The effects that pathogens have on the body, such as a <u>fever</u> (raised body temperature) or a <u>rash</u>, are called <u>symptoms</u>. Different microorganisms cause different symptoms, but they all damage the body in one way or another. The damage is done to the body's <u>cells</u>.

2) Some microorganisms damage cells <u>directly</u>.

3) Many microorganisms produce <u>poisonous chemicals</u> called <u>toxins</u> that damage cells.

4) The toxins made by many bacteria poison cells, causing <u>fever</u> or <u>inflammation</u> (painful swelling). Some strains of <u>Escherichia coli</u> cause <u>diarrhoea</u> by secreting toxic substances.

Your Body Has Barriers to Keep Microorganisms Out

Fortunately, it's <u>not</u> very easy for microorganisms to get into your body and start causing havoc. Your body has some pretty nifty <u>defence systems</u> in place:

1) The <u>skin</u> is a really effective barrier for keeping microorganisms out. And if the skin is damaged, it can rapidly <u>repair</u> itself to stop wounds from getting infected.

2) The skin does contain tiny openings so that <u>sweat</u> can come out to cool you down. Luckily sweat contains substances that reduce the growth of microorganisms. It forms an extra <u>protective layer</u> over the skin.

3) Your eyes produce <u>tears</u>, which contain chemicals that can <u>kill</u> bacteria.

4) Bacteria that enter your body in food or drink usually get killed by the <u>acid</u> in your stomach. This stops them from spreading through your body.

Don't wash — fight disease using sweat alone (yummy)...

The key things that you must get into your head from this page are:
Microorganisms <u>spread rapidly</u>, they cause symptoms by <u>damaging cells</u>, and your body has ways of <u>keeping them out</u>. Make sure you've learnt all the little details too and you'll be fine.

The Immune System

From time to time microorganisms <u>do</u> make it past your outer defences and into the body. But all is not lost, because your body still has a pretty powerful weapon left — your <u>white blood cells</u>.

Your Immune System Fights Off Invading Microorganisms

The role of the <u>immune system</u> is to deal with any microorganisms that enter the body. An immune response <u>always</u> involves <u>white blood cells</u>. There are <u>different types</u> that have different jobs to do.

1) One type of white blood cell is able to detect things that are '<u>foreign</u>' to the body, e.g. microorganisms.

2) These white blood cells <u>engulf</u> the microorganisms and <u>digest them</u>.

3) They're <u>non-specific</u> — they attack <u>anything</u> that's not meant to be there.

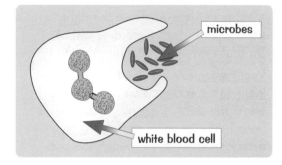

Antibodies Recognise Microorganisms

A different group of white blood cells attack <u>specific</u> microorganisms.

1) Every invading microorganism has unique molecules (called <u>antigens</u>) on its surface.

2) When your white blood cells come across a <u>foreign antigen</u> (i.e. one they don't recognise), they start to produce molecules called <u>antibodies</u> which lock on to and kill the invading pathogens.

3) The antibodies produced are <u>specific</u> to that type of antigen — they won't lock on to any others.

4) Once a white blood cell recognises the antigens on a microorganism, it <u>divides rapidly</u> to make more identical cells, which make lots of the right <u>antibody</u> to get on with fighting the infection.

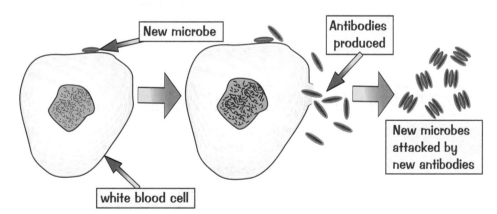

5) Some white blood cells stay around in the blood after the original infection has been fought off. They can reproduce very quickly if the <u>same</u> antigen enters the body for a <u>second</u> time.

6) That's why you're immune to <u>most</u> diseases that you've already had — the body carries a '<u>memory</u>' of what the antigen was like, and can <u>quickly produce</u> loads of <u>antibodies</u> if you get infected again.

White blood cells — like Rambo, but smaller. And rounder...

Have you ever noticed that loads of people seem to get <u>colds</u> as soon as they come back to school after the <u>summer holidays</u>? The reason is that people have been away on <u>holiday</u> and brought back new microorganisms that most other people don't have the <u>antibodies</u> for.

Vaccination

Some diseases can be pretty nasty, so if possible you want to give your immune system a <u>head start</u> in fighting the microorganisms that cause it. You can do this using a <u>vaccine</u>.

Vaccinations Use a <u>Safe</u> Version of a <u>Dangerous</u> Microorganism

1) When you're infected with a new <u>microorganism</u>, it takes your white blood cells a few days to get their numbers up and to make the right antibodies to help them deal with it. By that time, you can be pretty <u>ill</u>.

2) <u>Vaccination</u> involves injecting <u>dead</u> or <u>inactive</u> microorganisms. These still carry the same <u>antigens</u>, which means your body produces <u>antibodies</u> to attack them — even though the microorganism is <u>harmless</u> (since it's dead or inactive).

3) For example, the MMR vaccine contains <u>weakened</u> versions of the viruses that cause <u>measles</u>, <u>mumps</u> and <u>rubella</u> (German measles) all together.

4) If live microorganisms of the same type appear after that, the white blood cells <u>rapidly</u> produce antibodies to kill off the infection.

5) This normally means you can get rid of all the disease-causing microorganisms <u>before</u> they reach a level that makes you ill.

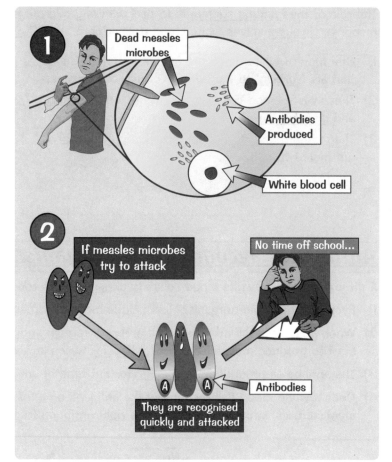

Some Microorganisms <u>Change Very Quickly</u>

1) When some microorganism <u>reproduce</u> they undergo <u>random changes</u>, this means the next generation is <u>different</u> from the previous one.

2) It's hard to develop <u>vaccines</u> for these kinds of microorganism, because the changes can lead to them having different <u>antigens</u>.

3) If you've been vaccinated against <u>one strain</u> of the microorganism, and then it changes its antigens, your immune system has to recognise the <u>new antigens</u>.

4) The virus that causes <u>influenza</u> (flu) is one kind of microorganism that changes <u>very quickly</u>. <u>New vaccines</u> against new strains of influenza have to be developed on a regular basis.

The first ever vaccine was invented by a cow...

Well, sort of. <u>Edward Jenner</u> is usually credited with giving the first vaccine in <u>1796</u>. He infected a boy with <u>cowpox</u>, a cow disease that causes only mild symptoms in humans. The antibodies that the boy's immune system produced to fight off cowpox were <u>so similar</u> to antibodies for <u>smallpox</u> that the boy became <u>immune</u> to that, too. (But Jenner only got the idea from noticing that cow farmers rarely got smallpox — because they'd already caught cowpox — so I think we should just credit the <u>cows</u>.)

Vaccination — Pros and Cons

Even if you <u>can</u> give someone a vaccine to prevent disease, not everyone thinks you necessarily <u>should</u>...

There are <u>Advantages</u> and <u>Disadvantages</u> of Vaccination

ADVANTAGES

1) It <u>prevents disease</u> — a pretty big advantage I reckon.
2) Big outbreaks of disease can be prevented if a <u>large percentage</u> of the population is vaccinated.
3) Some diseases, such as <u>smallpox</u>, have been virtually <u>wiped out</u> by vaccination programmes.

DISADVANTAGES

1) Vaccinations can <u>never</u> be <u>completely safe</u> for everyone, because individuals will have varying degrees of <u>side-effects</u> from a vaccine.
2) For example, one in four children who have the <u>meningitis vaccination</u> develop a painful <u>swelling</u> at the site of the injection, and one in 50 have a <u>fever</u> after they're given the vaccine.

Individuals <u>May Not Want to Do What's</u> <u>Best</u> <u>for</u> <u>Society</u>

1) Some people are so worried about the possible side effects of some vaccines that they <u>refuse</u> to be vaccinated, or won't allow <u>their children</u> to be vaccinated.
2) There must be enough people in a population who <u>are</u> vaccinated to <u>control</u> the disease.
3) These <u>conflicting interests</u> mean it can be hard to agree on a vaccination policy.

4) Some people think that the right policy is the one that leads to the <u>best outcome</u> for the <u>majority</u> of people in the population. This would probably mean vaccinating <u>everyone</u> — the number of people who suffer <u>serious</u> side effects is likely to be far <u>lower</u> than the number of people who would suffer serious effects from diseases like measles and rubella otherwise.

5) Another argument is that it's <u>unfair</u> for some people to avoid all risk by refusing to be vaccinated, because they still <u>benefit</u> from living in a society where most people <u>are</u> vaccinated. They're unlikely to catch the disease because there's no-one to pass it on to them — everyone else has risked having the vaccine.

6) The reverse argument is that nobody should be <u>forced</u> to have a vaccination that they don't want — each person should have the <u>right</u> to decide about vaccination for themselves and their children.

7) Some people believe that certain actions (e.g. injecting something into the body) are <u>never</u> justified, even if they do benefit the majority, because they're unnatural or wrong — but not everyone agrees on what these actions are.

"It won't hurt"
— yeah right.

Come on, let meee–eeeeee... vaccinate yoooouu...

So vaccination really comes down to <u>personal choice</u> and what you believe is <u>fair</u>. It's important that you understand the <u>different arguments</u> for and against vaccination, and think about the <u>risks</u> associated with it too. Remember, vaccines are <u>never</u> totally safe, but then <u>what is</u>? I reckon they're a better option than getting a load of <u>grim</u> diseases anyway...

Antibiotics

The discovery of the first <u>antibiotic</u>, penicillin, was a huge one in medicine — suddenly infections that had often been fatal could be <u>cured</u>. But unfortunately antibiotics may <u>not</u> be a permanent solution.

Antibiotics Can Kill Bacteria and Fungi, but Not Viruses

1) <u>Antibiotics</u> are drugs that can kill <u>bacteria</u> and <u>fungi</u> without damaging your own body cells.

2) They're very useful for clearing up infections that your own immune system is having <u>trouble</u> with.

3) However, they <u>don't kill viruses</u>. <u>Flu and colds</u> are caused by <u>viruses</u> and basically you just have to <u>wait</u> for your body to deal with them and <u>suffer</u> in the meantime. <u>AIDS</u> is also caused by a virus (HIV), which is why it can't be cured with antibiotics.

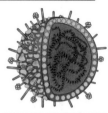

A horrid Flu Virus

Bacteria Can Evolve and Become Antibiotic-Resistant...

1) <u>Random changes</u> sometimes take place in a population of bacteria, so that one or two individuals have different characteristics from the rest.

2) Sometimes these changes mean that those individuals are <u>less vulnerable</u> to antibiotics.

3) An antibiotic will then kill off all the <u>non-resistant bacteria</u>, but leave the <u>resistant strain</u> to <u>multiply</u> and <u>spread</u>.

4) This is a problem for people who become infected with these microorganisms, because you <u>can't</u> easily get rid of them with antibiotics. Sometimes drug companies can come up with a <u>new</u> antibiotic that's effective, but '<u>superbugs</u>' that are resistant to most known antibiotics (e.g. MRSA) are becoming more common.

...So Do Everyone a Favour and Always Finish Your Antibiotics

1) The <u>more</u> antibiotics are used, the <u>bigger</u> the problem of antibiotic-resistance becomes. Although the antibiotic might get rid of most of the bacteria or fungi, there's always the danger that a few <u>resistant</u> individuals will be left behind.

2) If there are too many of these individuals left for your immune system to get rid of, they can then <u>reproduce</u> to create a <u>whole new population</u> of antibiotic-resistant microorganisms.

3) This is why it's important you take <u>all</u> the antibiotics a doctor prescribes for you. Lots of people stop bothering to take their antibiotics as soon as they <u>feel better</u>, but that just means that <u>most</u> of the microorganisms are gone. There could still be some <u>hardcore</u> ones lingering on.

4) It's also important that people only use antibiotics when they <u>really</u> need to. It's not that antibiotics actually <u>cause</u> resistance, but they do create a situation where naturally resistant microorganisms have an <u>advantage</u> and so increase in numbers. If they're not actually doing you any good, it's pointless to take antibiotics — and it could actually be harmful for everyone else.

Aaargh, a giant earwig! Run from the attack of the superbug...

The reality of <u>superbugs</u> is possibly even scarier than giant earwigs. Actually, nothing's more scary than giant earwigs, but microorganisms that are <u>resistant</u> to <u>all</u> our drugs are a worrying thought. It'll be like going <u>back in time</u> to before antibiotics were invented. So far <u>new drugs</u> have kept us one step ahead, but some people think it's only a matter of time until the options run out.

Drug Trials

You can't just get any old chemical and market it as a new wonder drug. Any new drug has to go through loads of tests to make sure it's safe to use, and also to make sure it actually does what you claim it does.

Drugs are Tested First in a Laboratory

1) New drugs are being developed all the time to help fight different diseases.

2) These drugs are often developed using human cells that are grown in the lab. This means that you can measure the effect the drug has on real human cells, rather than on another species' cells.

3) On the other hand, it can't recreate the conditions of a whole system or organism, so you still can't be sure that the drug's safe to use, or that it actually works.

4) To make sure, all new drug must be tested on at least two different species of live mammal (rats and monkeys are often used) before it's given to humans. This way, potentially harmful substances are usually weeded out before the drugs are given to human volunteers.

5) Many mammals have systems that are similar to those of humans, so the tests give early indications of what the drug might do in the human body.

6) If the drug causes serious problems in the animals, the testing is unlikely to go any further, and this saves any humans from being harmed.

Drugs are Then Tested on Humans in Clinical Trials

1) If the laboratory tests don't pick up on any possible bad side-effects from the drug, it will then go on to be tested on human volunteers.

2) These tests are called clinical trials...

First the Drug is Tested on Healthy Volunteers...

This is to make sure it doesn't have any harmful side effects when the body is working normally. Sick people are likely to be more vulnerable to any damage the drug could do, which is why the drug isn't tested on them yet.

...Then on Actual Sufferers

If the results of the tests on healthy volunteers are good, the drugs can be tested on people suffering from the illness. These are tests for both safety and effectiveness — this is where you find out whether or not the drug actually works.

This page might be a bit testing but you need to know it...

You can't know for sure what will happen in a complete human system until you test one. Lab tests are designed to make the drug as safe as possible before clinical trials start, but there have been times when totally unexpected side-effects have occurred, some of them serious. Having said that, if nobody ever took part in trials, there would never be any new drugs.

The Circulatory System

Blood is vital. It pumps all sorts of stuff around your body, from oxygen and carbon dioxide to the useful bits of last night's pizza. And your heart and blood vessels are pretty well adapted for the task...

The Heart and Blood Vessels Supply Blood to the Body

1) Blood is circulated around the body in tubes called blood vessels. Oxygen and food are carried in the blood to the body cells, and waste substances such as carbon dioxide are carried away from the cells.

2) The heart is a pumping organ that keeps the blood flowing through the vessels. It's made up of muscle cells that keep it beating continually. These cells have their own blood supply to deliver the food and oxygen needed to keep them working properly.

3) The guy with no arms is helpfully displaying how the heart is connected to everything in your body by a network of blood vessels.

There are Two Major Types of Blood Vessel

1) Arteries carry blood that's high in oxygen away from the heart to the body cells (including the heart muscle).

2) It comes out of the heart at high pressure, so the artery walls have to be strong and elastic. They're much thicker than the walls of veins...

elastic fibres and smooth muscle

lumen

brain

lungs

heart

liver

gut

kidneys

from lower limbs

to lower limbs

1) Veins carry blood that's low in oxygen back to the heart.

2) The blood is at a lower pressure in the veins so the walls don't need to be as thick.

3) They have a bigger lumen (the hole down the middle) than arteries, to help the blood flow more easily.

4) They also have valves to help keep the blood flowing in the right direction.

large lumen

elastic fibres and smooth muscle

valves

Fatty Deposits Can Cause Problems for the Heart and Brain

1) Cholesterol is a fatty substance which can build up on the walls of your arteries.

2) This makes the lumen of the arteries narrower.

3) The narrower lumen restricts the flow of blood — bad news for the part of the body the affected artery is supplying with food and oxygen.

4) If an artery supplying the heart muscles or brain is affected, it can cause a heart attack or stroke.

5) A diet high in saturated fat can increase the amount of cholesterol in your blood vessels.

6) If you swap saturated fat for unsaturated fat, the amount of cholesterol can decrease.

Unbreak my heart — say it's pumping again...

So in fact, your boyfriend or girlfriend saying they don't want to go out with you any more won't really break your heart, despite what all these warblers on the radio seem to think. Putting on eight stone in weight through comforting yourself with ice cream might do it though.

Heart Disease

You know from earlier in the section that loads of <u>diseases</u> are caused by nasty <u>microorganisms</u> that sneak into the body. But this <u>isn't</u> usually the case with <u>heart disease</u>.

Your Lifestyle Can Increase Your Risk of Heart Disease

1) Heart disease is linked to <u>lifestyle factors</u>, e.g. what someone <u>eats</u> and how much <u>exercise</u> they do. Some people might be more at <u>risk</u> because of their <u>genes</u> too. In most people it's caused by one or both of these things, rather than by microorganisms.

2) The main lifestyle factors that increase the risk of heart disease are a <u>poor diet</u>, <u>stress</u>, <u>smoking</u>, and too much <u>alcohol</u>.

3) Heart disease is more common in <u>industrialised countries</u>, such as the UK and USA, than in non-industrialised countries. This is mainly because people in these countries can <u>afford</u> a lot of food that's high in saturated fat and often don't <u>need</u> to be very <u>physically active</u>.

bad bad bad

good good good

4) <u>Taking regular moderate exercise</u> reduces the risk of developing heart disease. This is because exercise <u>burns fat</u>, preventing it building up in the arteries. Exercise also <u>strengthens</u> the heart muscle.

Epidemiological Studies Can Identify Possible Risk Factors

<u>Epidemiology</u> is the study of <u>patterns</u> of disease, the factors that affect the <u>spread</u> of disease, and why some groups of people are more <u>likely</u> to get certain illnesses than others.

1) Studies into the <u>epidemiology</u> of <u>heart disease</u> have shown which factors are linked to the disease.

2) People who <u>smoke</u> more have been shown to suffer more from heart disease, so there's said to be a <u>correlation</u> between smoking and heart disease (see next page).

3) As with all scientific studies, epidemiological studies are more convincing if they're based on <u>lots</u> of cases. Single cases <u>don't</u> provide good evidence for or against a correlation.

4) Scientific research can only be considered <u>trustworthy</u> once it's been <u>peer reviewed</u>. This means that it's evaluated by <u>other scientists</u> (the researcher's peers) before the claims are widely circulated. The methods and results are then published in <u>scientific journals</u>, so that anyone can read them and make up their own mind. Some scientists might decide to carry out <u>similar studies</u> to see if they get the <u>same results</u>, and some might come up with <u>alternative explanations</u> for the results.

5) If loads of different scientists carry out similar studies and <u>find the same thing</u>, then the new scientific claim is probably <u>pretty reliable</u>. But if there's <u>absence of replication</u> — i.e. people have investigated the same thing, but <u>don't agree</u> on the results — then people might <u>doubt</u> the new claim.

6) It usually takes a lot of discussion, argument and further studies before a new idea becomes accepted. This is generally a good thing — there are often lots of <u>different</u> ways to explain the data, and even if someone's done the research it doesn't mean they'll pick the <u>right</u> one.

Usually it's peer review, occasionally it's handbags at dawn...

Scientists through the ages have come up with some <u>amazing theories</u>, but every so often they act like great big kids — <u>bickering</u> over someone else's results because they had the same idea but <u>got there too late</u>. Ah well, at least it makes sure that <u>rubbish ideas</u> aren't automatically accepted as true.

Heart Disease — Correlation and Cause

It takes a long time for any new scientific discovery to become accepted. And here's why...

Statistical Correlations Don't Prove One Thing Causes Another

1) A correlation is basically just a relationship between two factors.

For example, say scientists monitored the number of heart attacks in smokers. If they got results like those shown on the right, this would indicate that there's a correlation between these two factors — the more cigarettes the people in this study group smoked per day, the more likely they were to have a heart attack.

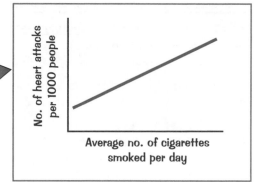

2) Many scientific studies aim to work out whether there's a correlation between two factors. But you have to be careful, because a correlation doesn't necessarily mean that one thing causes the other.

Scientific Studies Must Be Carefully Planned

A lot of different things can affect how valid the results of a study are. And I mean a whole lot of things...

1) When studying the relationship between two variables, scientists need to consider all the other things that could be contributing to the outcome they're interested in.

For example, the study looking at how smoking affects the risk of heart disease would need to take into account many other factors too — things like the body weight of the people in the study and the amount of exercise they do could also affect their risk of heart disease.

Studies like this must try to minimise the effects of these other factors. The study should use volunteers who have similar body weights and who do similar levels of exercise.

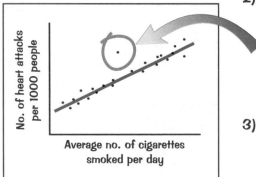

2) A study into the effect of smoking on heart disease should also look at as many people as possible. This reduces the chance of any unusual results (called anomalies) affecting the overall pattern. The more people (or lab rats, or bananas, or repetitions of the experiment) there are in a study, the more reliable it's likely to be.

3) This is why, when people say things like "My gran smoked 60 cigarettes a day from the age of 12 and lived to be 90," it doesn't actually mean cigarettes are harmless. There are always going to be unusual results that happen by chance. One individual case doesn't provide evidence for or against anything — you need a big sample to be able to show a significant correlation.

I think there's a correlation between high BP and this page...

Oh how examiners love to talk about correlations. It's fairly easy really, all you have to do is break it down into factors (e.g. smoking or a high cholesterol diet) and outcomes — the things the factors cause.

Revision Summary for Module B2

Another section, another revision summary, in much the same way that night follows day. Now, as you answer these questions you can take part in a little experiment. My theory is that the number of questions you can answer correctly without looking back through the section depends on your star sign. Have a go and then see if your score matches my predictions at the bottom of the page. (Which are based on the alignment of the moon with Jupiter and whether or not I like the sound of the star sign name.)

1) What name is given to microorganisms that cause disease?
2) How do microorganisms cause disease symptoms?
3) Give three different ways in which your body is adapted to keep microorganisms out.
4) Explain how white blood cells help to get rid of microorganisms that do enter the body.
5) Explain why you are immune to most diseases that you've already had.
6)* Vaccination involves injecting inactive microbes. How does this give you immunity in the future?
7) Which three diseases can be prevented by the MMR vaccine?
8) Vaccines against the flu are different every year. Why is this?
9) What is an antibiotic?
10) Explain why antibiotics should not be prescribed for someone with the flu.
11) What is a 'superbug'?
12) Why are patients advised to complete a course of antibiotics even if they start to feel better?
13) Give two ways that new drugs are usually tested before they're given to humans.
14) Why are new drugs tested on healthy people first, not patients with the illness they're designed to cure?
15) How do the heart's muscle cells get food and oxygen?
16) How is the structure of an artery adapted to its function?
17) How is the structure of a vein adapted to its function?
18) Explain how a diet high in saturated fat can cause a heart attack.
19) Your lifestyle can increase your risk of heart disease. What else can increase your risk?
20) Give the four main lifestyle factors associated with an increased risk of heart disease.
21) Why is heart disease more common in industrialised countries?
22) Explain how peer review increases the trustworthiness of scientific studies.
23) How do scientists minimise the effects of factors other than those they're studying?
24) Why should an epidemiological study include as many people as possible?

PREDICTIONS: Aries 15–20. Taurus 14–18. Gemini 24 or more. Cancer 10–15. Leo 13–18. Virgo 10–14. Libra 20–24. Scorpio 24 or more. Sagittarius 14–18. Capricorn 17–22. Aquarius 24 or more. Pisces — you haven't even tried them yet! Get on with it!

* Answer on page 100.

Natural and Synthetic Materials

This section is all about <u>materials</u>, their <u>properties</u> and the best ones to use to make different <u>products</u>.

All Materials are Made Up of Chemicals

Absolutely everything is made up of <u>chemicals</u>. The materials we use are either made of individual <u>chemicals</u> or <u>mixtures</u> of chemicals.

Chemicals are made up of <u>atoms</u> or <u>groups of atoms</u> bonded together.

1) Iron is a <u>chemical</u> <u>element</u> — it's made up of <u>iron atoms</u>. **Fe**

2) Water is a <u>chemical</u>. It's made up of lots of <u>water molecules</u>. A molecule is a group of atoms <u>bonded</u> together. Water molecules contain <u>2 chemical elements</u> — hydrogen and oxygen. **(H) (O) (H)**

Some materials are <u>mixtures</u> of chemicals. A mixture contains different substances that are <u>not</u> chemically bonded together. For example, <u>rock salt</u> is a mixture of two compounds — salt and sand.

Some of These Materials Occur Naturally...

A lot of the materials that we use are made by other <u>living things</u>, not by humans:

MATERIALS FROM PLANTS

1) <u>Wood</u> and <u>paper</u> are both made from <u>trees</u>.
2) <u>Cotton</u> comes from the cotton plant.

MATERIALS FROM ANIMALS

1) <u>Wool</u> comes from <u>sheep</u>.
2) <u>Silk</u> is made by the <u>silkworm</u> larva.
3) <u>Leather</u> comes from <u>cows</u>.

...Others are Synthetic — Made by Humans

We often use <u>man-made</u> (synthetic) materials instead of <u>natural</u> materials, e.g.

1) All <u>rubber</u> used to come from the sap of the <u>rubber tree</u>. We <u>still</u> get a lot of rubber this way (e.g. for car tyres), but you can also make rubber in a <u>factory</u>. The advantage of this is that you can <u>control</u> its <u>properties</u>, making it suitable for different <u>purposes</u>, e.g. wetsuits. However, it is <u>cheaper</u> to use rubber from the sap of the rubber tree.

2) A lot of <u>clothes</u> are made of <u>man-made</u> fabrics like <u>nylon</u> or <u>polyester</u>. These materials are often a lot <u>cheaper</u> and have more <u>uses</u> than natural fabrics like wool and silk — e.g. you can make fabrics that are super-stretchy, or sparkly.

3) Most <u>paints</u> are mixtures of man-made chemicals. The <u>pigment</u> (the colouring) and the stuff that holds it all together are designed to be <u>tough</u> and to stop the colour fading.

So silk comes out of a worm's bottom then...

OK, so now you should know what a <u>material</u> is (and that it doesn't just mean <u>fabric</u>). You should also have a pretty good idea of the different kinds of materials around, and where they come from. Not bad for the first page of the section. Time to cover the page and <u>scribble</u> down what you can remember.

Materials and Properties

Not all materials are the same, as you'll find out if you try to make a hat out of spaghetti hoops...

Different Materials Have Different Properties

MELTING POINT

Most materials that are pure chemicals have a unique melting point. This is the temperature where the solid material turns to liquid. E.g. the melting point of water is 0 degrees Celsius (0 °C).

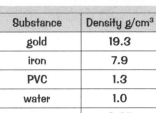

STRENGTH

Strength is how good a material is at resisting a force. You can judge how strong it is by how much force is needed to either break it or permanently change its shape (deform it). There are two types of strength you need to know about:

1) TENSILE (OR TENSION) STRENGTH — how much a material can resist a pulling force. Things like ropes and cables need a high tensile strength, or they'd snap.

2) COMPRESSIVE STRENGTH — how much a material can resist a pushing force. Building materials like bricks need good compressive strength, or they'd be squashed by the weight of the bricks above them.

Some things, like cross beams in roofs, need to be made of materials with good compressive and tensile strength — they get both pushed and pulled.

STIFFNESS

A stiff material is good at not bending when a force is applied to it. This isn't the same as strength — a bendy material can still be strong if a big force doesn't permanently deform it.

1) Materials like steel are very difficult to bend — they're very stiff.

2) Some kinds of rubber are very strong but they bend and stretch very easily — they're not stiff.

HARDNESS

The hardness of a material is how difficult it is to cut into.

1) The hardest material found in nature is diamond.

2) The only material that can cut a diamond is another diamond.

3) Diamonds can cut most other materials — many industrial drills have diamond tips.

DENSITY

Density is a material's mass per unit volume (e.g. g/cm^3). Don't confuse density with mass or weight.

1) Air is not very dense. You'd need a huge volume of it to make up 1 kg in mass.

2) Gold is very dense. A small volume of gold would make up 1 kg in mass.

3) Objects that are less dense than water will float (like ice). Objects that are more dense than water will sink.

Substance	Density g/cm^3
gold	19.3
iron	7.9
PVC	1.3
water	1.0
ice	0.97
air	0.001

'Cos diamonds are an industrial drill's best friend...

Learn everything on this page — I make that five things. And make sure you're clear about what density is and why it's **NOT** the same as mass. Measure out 1 kg of loose change (or rocks) and 1 kg of breakfast cereal and compare the different volumes — same mass, but very different densities...

Making Measurements

When you measure the properties of a material you can only accept the results as reliable if other scientists elsewhere can get similar results — your experiment has to be <u>repeatable</u>.

Most Measurements Involve a Degree of Uncertainty

When you're <u>measuring something</u> there are loads of reasons why your results might <u>not</u> be accurate:

1) There might be a <u>fault</u> in your equipment.

2) Wrong results can be because of <u>human error</u> — <u>inaccurate</u> measuring, reading or recording of the results.

3) Your <u>samples</u> and <u>techniques</u> have to be the <u>same</u> every time. If lots of scientists all test the strength of iron they might all get <u>different results</u> because iron's strength depends on things like <u>how it's made</u>. The scientists need to work on <u>identical</u> samples using <u>identical techniques</u>.

Measurements Will Always Vary to Some Extent

If you want an <u>accurate result</u> you've got to take measurements <u>several times</u>. This is in case of a fault in your equipment or human error, etc. You won't get exactly the same measurement each time, but that's normal. Again, this could be down to faults in your equipment or human error.

Test	Density of gold g/cm³
A	19.3
B	19.4
C	12.8
D	19.1
E	30.1
F	19.2
G	19.5
H	19.2
I	19.2
J	19.5

This table shows the results of an experiment to measure the <u>density of gold</u> — the measurement was repeated 10 times:

1) The results of <u>tests C and E</u> are so <u>different</u> from the others that something must have gone <u>wrong</u>. Results like these that are an <u>abnormal distance</u> from the rest of the data are called <u>outliers</u>, and you can often just <u>ignore</u> them.

2) Plotting the results on a <u>graph</u> can help with spotting outliers — you can draw a '<u>line of best fit</u>' and ignore the results that aren't near it.

3) You <u>can't</u> always ignore outliers — if you're expecting the data to <u>vary</u> a lot, e.g. if you're measuring the height of children aged 1 to 10, then you <u>wouldn't</u> ignore a very low or high result.

4) The table suggests that the <u>true value</u> of the density of gold is in the <u>range</u> of 19.1 – 19.5 g/cm³, where most of the measurements lie.

5) A <u>mean</u> (average) gives you the <u>best estimate</u> of the <u>true</u> value. Take a <u>mean</u> of the remaining results (<u>add</u> them together then <u>divide</u> by the number of <u>good results</u> (8)). This works out at <u>19.3 g/cm³</u>.

Experiments Need to be Carefully Designed

1) Measuring a <u>mean</u> and ignoring <u>outliers</u> isn't enough. You also need to make sure that your experiment is a <u>fair test</u>.

2) The best way to make it a fair test is to vary only <u>one factor</u> in your experiment, and to only measure one thing at a time.

3) So if you're measuring densities of different materials, the <u>only thing</u> that you should vary each time is the material — that's the factor that you change. The volume of material that you test, the temperature and equipment that you use etc. must all be exactly the same each time.

4) Each time you <u>repeat</u> the test the other factors must be exactly the same — they must be <u>controlled</u>.

Designer experiment — mine's by Gucci darling...

This page is mostly about making sure that <u>measurements</u> are accurate, but it applies to <u>any</u> kind of scientific test really. If you don't follow these rules, then you <u>can't</u> really conclude anything from your results. These rules have to be followed by everyone, even the top scientists — so learn them now.

Materials, Properties and Uses

Every material has a <u>different</u> set of properties, which makes it <u>perfect</u> for some jobs, and totally <u>useless</u> for others. That probably explains why <u>chocolate teapots</u> have never really caught on...

The Possible Uses for a Material Depend on Its Properties

When you're choosing a material to use in a <u>product</u>, you need to think about its <u>properties</u>, e.g.

PLASTICS
- Can be fairly <u>hard</u>, <u>strong</u> and <u>stiff</u>
- Some are fairly <u>low density</u> (good for lightweight goods)
- Some are <u>mouldable</u> (easily made into things)

E.g. cases for televisions, computers and kettles

RUBBER
- <u>Strong</u> but soft and <u>flexible</u>
- <u>Mouldable</u>

E.g. rubber car tyres

NYLON FIBRES
- Soft and flexible
- Good <u>tensile strength</u>

E.g. ropes and clothing fabric

A Product's Properties Depend on the Materials It's Made From

Some products can be made from a <u>variety</u> of materials. How <u>good</u> the product is and how long it <u>lasts</u> depends on the <u>properties</u> of the materials it's made from.

1) <u>Gramophone records</u> 100 years ago were made of a <u>mixture</u> of natural materials like paper, slate and wax. There aren't many of these records left because they <u>broke</u> very easily — they weren't <u>strong</u>.

2) More <u>modern</u> records are made of <u>polyvinyl chloride (PVC)</u> or 'vinyl'. This material is <u>strong</u> and <u>flexible</u> so it's less likely to break. DJs sometimes still use vinyl records in clubs.

3) Most people these days own <u>compact discs (CDs)</u>. These are made of a very <u>tough</u>, <u>flexible</u> plastic called <u>polycarbonate</u>. It's quite strong and hard (it's used in bulletproof glass) and should last <u>even longer</u> than PVC — but we'll have to wait a while to find out...

You'll Need to Assess the Suitability of Different Materials

You need to be able to look at the <u>properties</u> of a material and work out what sort of <u>purposes</u> it might be <u>suitable</u> for, e.g.

1) <u>Cooking utensils</u> must be made from something with a <u>high melting point</u> that's <u>non-toxic</u>.

2) Material to make a <u>toy car</u> must be <u>non-toxic</u> and should be <u>strong</u>, <u>stiff</u> and <u>low density</u> — e.g. some kinds of plastic.

3) <u>Clothing fabric</u> mustn't be stiff, but needs a <u>good tensile strength</u> (so it can be made into fibres) and high <u>flame-resistance</u>, especially if it's for nightwear or children's clothes.

It's not rocket science — that's 'cos it's materials science...

It's all fairly <u>straightforward</u> stuff on this page — just be prepared to look at the properties of 'mystery' materials in the exam, and work out which material would be <u>most suitable</u> for different jobs. Don't think that any of the materials are totally useless — their properties will probably be <u>ideal</u> for something.

Chemical Synthesis and Polymerisation

Crude oil is formed from the buried remains of plants and animals — it's a fossil fuel. Over millions of years, with high temperature and pressure, the remains turn to crude oil, which can be drilled up.

Crude Oil is a Mixture of Lots of Substances

Crude oil is a mixture of hydrocarbons — molecules which are made of chains of carbon and hydrogen atoms only (see p.17 for more info). These chains are of varying lengths.

Crude Oil is Used to Make Loads of Synthetic Substances

1) Most of the hydrocarbons in crude oil are refined by the petrochemical industry to produce fuels and lubricants.

2) Only a very small amount of hydrocarbons from crude oil are chemically modified to make new compounds for use in things like plastics, medicines, fertilisers and even food.

Don't forget what synthetic means — it's man-made.

Polymerisation Means Loads of Small Molecules Link Together

Polymers are among the most important man-made materials — you can make loads of useful stuff from them — well, as long as you want a plastic-type thing, anyway. And I often do.

1) Plastics are formed when lots of small molecules join together to give a polymer.

2) They're usually carbon-based.

3) Under high pressure many small molecules 'join hands' (polymerise) to form long chains called polymers.

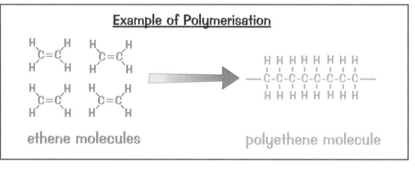

Example of Polymerisation

ethene molecules → polyethene molecule

Different Polymers Have Different Properties

Different polymers have different physical properties — some are stronger, some are stretchier, some are more easily moulded, and so on. These different physical properties make them suited for different uses.

- Strong, rigid polymers such as high-density polyethene are used to make plastic milk bottles.

- Light, stretchable polymers such as low-density polyethene are used for plastic bags and squeezy bottles. Low-density polyethene has a low melting point, so it's no good for anything that'll get very hot.

- PVC is strong and durable, and it can be made either rigid or stretchy. The rigid kind is used to make window frames and piping. The stretchy kind is used to make synthetic leather.

- Polystyrene foam is used in packaging to protect breakable things, and it's used to make disposable coffee cups (the trapped air in the foam makes it a brilliant thermal insulator).

- Heat-resistant polymers such as melamine resin and polypropene are used to make plastic kettles.

Revision — it's all about stringing lots of facts together...

If you're making a product, you need to pick your plastic carefully. It's no good making a kettle from plastic that melts at 50 °C — you'll end up with a messy kitchen, a burnt hand and no cuppa. You'd also have a bit of difficulty trying to wear clothes made of brittle, unbendy plastic.

Structures and Properties of Polymers

You need to know how the <u>properties</u> of a polymer are affected by the <u>way it's made</u>.

A Polymer's Properties Decide Its Uses

Its Properties Depend on How the Molecules are Arranged...

A polymer's properties don't just depend on the <u>chemicals</u> it's made from (see previous page).
The way the polymer chains are <u>arranged</u> has a lot to do with them too:

> If the polymer chains are packed close together, the material will have a high density.
> If the polymer chains are spread out, the material will have a low density.

...And How They're Held Together

The <u>forces</u> between the different chains of the polymer hold it together as a <u>solid mass</u>.

Weak Forces:
<u>Chains</u> held together by <u>weak forces</u> are
free to <u>slide</u> over each other. This means
the plastic can be <u>stretched easily</u>, and
will have a <u>low melting point</u>.

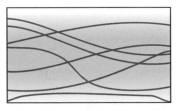

Strong Forces:
Plastics with <u>stronger bonds</u> between the polymer chains
have <u>higher melting points</u> and <u>can't be easily stretched</u>, as
the <u>crosslinks</u> hold the chains firmly together. Crosslinks are
<u>chemical bonds</u> between the polymer chains (see below).

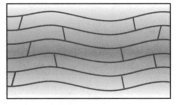

So, the <u>stronger</u> the bonds between the polymer chains, the more
<u>energy</u> is needed to break them apart, and the <u>higher</u> the <u>melting point</u>.

Polymers Can be Modified to Give Them Different Properties

You can <u>chemically modify</u> polymers to change their <u>properties</u>.

1) Polymers can be modified to <u>increase</u> their <u>chain length</u>. Polymers with <u>short</u> chains are <u>easy</u> to shape
and have <u>lower</u> melting points. <u>Longer</u> chain polymers are <u>stiffer</u> and have <u>higher</u> melting points.

2) Polymers can be made stronger by adding <u>cross-linking agents</u>.
These agents chemically <u>bond</u> the chains together, making the
polymer <u>stiffer</u>, <u>stronger</u> and more <u>heat-resistant</u>.

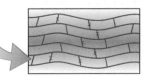

3) <u>Plasticisers</u> can be added to a polymer to make it <u>softer</u> and easier to shape.
Plasticisers work by getting in <u>between</u> the polymer chains and <u>reducing</u> the forces between them.

Choose your polymers wisely...

The molecules that make up a plastic affect the <u>properties</u> of the plastic, which affect what the plastic
can be used for. I know there are a lot of diagrams of lines on this page, but the <u>structure</u> of polymers
is really important — it affects their properties, and it might even come up in the exam.

Life Cycle Assessments

If a company wants to manufacture a new product, they carry out a <u>life cycle assessment (LCA)</u>. This looks at every <u>stage</u> of the product's life to assess the <u>impact</u> it would have on the environment.

Life Cycle Assessments *Show* Total Environmental Costs

A <u>life cycle assessment (LCA)</u> looks at each stage of the life cycle of a product, from the <u>raw materials</u> to when it's <u>disposed</u> of, and works out the potential <u>environmental impact</u>:

1) <u>Extracting and refining raw materials</u>

E.g. <u>metals</u> may have to be <u>mined</u> from the ground, then <u>extracted</u> from their ore. Polymers come from crude oil, which has to be <u>drilled</u> and <u>refined</u>. All these things use <u>energy</u>, which usually means burning <u>fossil fuels</u>. Sometimes there's a <u>limited supply</u> of the raw material, too — e.g. oil supplies may run out soon.

2) <u>Manufacturing the product</u>

This often uses a lot of <u>energy</u> and may cause <u>pollution</u> and use other resources — e.g. making a new car uses about 9000 litres of water.

3) <u>Using the product</u>

Just <u>using</u> the finished product can also damage the environment — e.g. an <u>electrical product</u> like a TV uses electricity likely to come from burning <u>fossil fuels</u>.

4) <u>Disposing of the product</u>

When people have finished with the product, it has to be disposed of. It could be <u>incinerated</u> (burnt), which might cause air pollution, or it could go into a <u>landfill site</u>, or be <u>recycled</u>. All these options have an <u>environmental impact</u>.

Life Cycle Assessments *are* Helpful for *Making Decisions*

An <u>LCA</u> helps you work out the best <u>materials</u> and <u>manufacturing process</u> for your product. If the <u>environmental impact</u> or the <u>cost</u> is too high, you can choose another material or manufacturing process. If it's quite low, you might decide to use the materials for <u>other products</u> too.

Companies can use the information from an LCA to set up a process which <u>doesn't</u> harm the environment so much that future generations suffer. This is a key principle of <u>sustainable development</u>.

An example of sustainable development would be a paper company only using wood from forests that are <u>replanted</u> and regrow <u>faster</u> than the company is felling them, and taking steps to <u>protect</u> the wildlife that lives in the forests.

Need exercise? Go life-cycling then...

The <u>environmental cost</u> of a product can vary a lot. If it's something that goes out of date quite quickly, like a computer, then it will have a short life cycle. However, high quality expensive furniture might be kept for a lifetime, and its long life cycle means it has less impact per year on the environment.

Revision Summary for Module C2

This section has gone from silk and rubber to crude oil and then onwards to environmental issues. That's an awful lot to take in in one section. Whether you find this topic easy or hard, interesting or dull, you've simply got to learn it all before the exam. Try these questions and see how much you really know:

1) Give one example of a chemical element.

2) Name a material we use that comes from:
   a) plants,   b) animals.

3) Why do we use both natural and synthetic rubber?

4) What's the difference between compressive and tensile strength?

5) What's the hardest material found in nature?

6) Give a definition of density.

7) You're measuring the properties of a material — give three reasons why your results might not be accurate.

8) Why should you always repeat your measurements?

9)* You do an experiment to measure the density of a mystery material. You do the measurements nine times, and your results (in g/cm^3) are as follows: 8.1, 8.3, 8.1, 8.0, 14.2, 8.3, 8.4, 8.0, 8.2.
   a) Which result would you discard from your data?
   b) What is the mean of your measurements of the density of the mystery material?

10) How can you ensure that an experiment is a fair test?

11) Name two properties of each of the following materials that make them useful in manufacturing:
   a) Plastic,   b) Rubber,   c) Nylon.

12)* Name three properties that you'd look for when choosing a material to make a child's dinner bowl.

13) What does crude oil contain?

14) Briefly describe what happens during polymerisation.

15) Give an example of a product where a polymer has replaced a natural material.

16) How does the arrangement of polymer chains affect the density of a material?

17) A polymer is easily stretched and has a low melting point. What can you say about the arrangement of its molecule chains and the forces holding them together?

18) What would you add to a polymer to make it stiffer and stronger?

19) How do plasticisers work?

20) Name the four main stages considered in a life cycle assessment.

21) Give an example of sustainable development.

* Answers on page 100.

Electromagnetic Radiation

Light, X-rays and microwaves are all the <u>same kind of thing</u>, just quite different. Right, that's clear then.

Light *is a Type of* Electromagnetic Radiation

1) <u>Light</u> is a type of <u>electromagnetic radiation</u> (EM radiation).

2) <u>Radiation</u> is just a <u>transfer of energy</u>. E.g. <u>sunlight</u> is a transfer of energy from the Sun to the Earth.

3) <u>Visible light</u> (the seven colours of the rainbow, from red to violet) is just radiation that our <u>eyes</u> can <u>detect</u>. Radiation that's further along in the 'red' direction is called <u>infrared</u>. Similarly, radiation that's further along in the 'violet' direction is called ultraviolet.

4) There are <u>seven</u> types of radiation altogether, making up the <u>electromagnetic spectrum</u> —

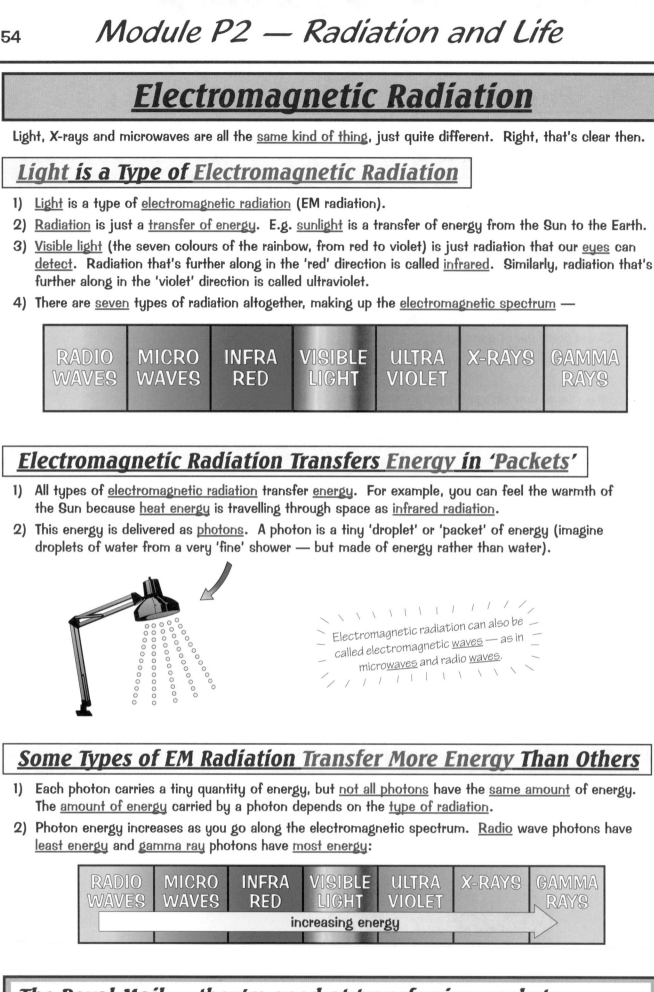

| RADIO WAVES | MICRO WAVES | INFRA RED | VISIBLE LIGHT | ULTRA VIOLET | X-RAYS | GAMMA RAYS |

Electromagnetic Radiation Transfers *Energy in 'Packets'*

1) All types of <u>electromagnetic radiation</u> transfer <u>energy</u>. For example, you can feel the warmth of the Sun because <u>heat energy</u> is travelling through space as <u>infrared radiation</u>.

2) This energy is delivered as <u>photons</u>. A photon is a tiny 'droplet' or 'packet' of energy (imagine droplets of water from a very 'fine' shower — but made of energy rather than water).

Electromagnetic radiation can also be called electromagnetic <u>waves</u> — as in micro<u>waves</u> and radio <u>waves</u>.

Some Types of EM Radiation *Transfer More Energy* Than Others

1) Each photon carries a tiny quantity of energy, but <u>not all photons</u> have the <u>same amount</u> of energy. The <u>amount of energy</u> carried by a photon depends on the <u>type of radiation</u>.

2) Photon energy increases as you go along the electromagnetic spectrum. <u>Radio</u> wave photons have <u>least energy</u> and <u>gamma ray</u> photons have <u>most energy</u>:

| RADIO WAVES | MICRO WAVES | INFRA RED | VISIBLE LIGHT | ULTRA VIOLET | X-RAYS | GAMMA RAYS |

increasing energy →

The Royal Mail — they're good at transferring packets...

Remember that EM radiation is just the <u>transfer of energy</u>, as <u>photons</u>. <u>Visible light</u> is just electromagnetic radiation that we can <u>see</u> — and very pretty it is too. There's nothing special about it though — other creatures 'see' other parts of the spectrum — bees see ultraviolet, for instance.

EM Radiation and Energy

All sorts of objects emit EM radiation. You're emitting some infrared radiation at this very moment.

EM Radiation is Emitted from a Source...

1) Many objects emit electromagnetic radiation, e.g. the Sun, radio transmitters, mobile phones, etc. Any object that emits radiation is called a source.
2) Once emitted, all types of EM radiation can travel through space (a vacuum). In a vacuum, all EM radiation travels at the same speed — the 'speed of light'.

...And Transmitted, Reflected or Absorbed Somewhere Else

1) When radiation is emitted from a source, it spreads out until it reaches some matter (a substance — like air, glass, walls...). Three things can then happen:

> • The radiation might be TRANSMITTED — just keep going, like light passing through glass.
>
> • It could be REFLECTED — bounce back, like light reflected from a mirror.
>
> • Or the radiation could be ABSORBED — like a sunbather absorbing UV rays from the Sun.

2) What happens depends on what the substance is like and the type of radiation.
3) Two or three of these things can happen at the same time. E.g. when sunlight shines on glass a lot of the light is transmitted, but some of it is reflected — so you can check your hair in shop windows.
4) Radiation can be absorbed by objects a long way from the source, e.g. when a parked car warms up in the sunshine:

Infrared radiation is emitted from the Sun and travels 150 000 000 km through space, to Earth. Some of it is then absorbed or reflected by the atmosphere — but some of it reaches the car and warms it up.

5) Objects that absorb and register radiation are called detectors — e.g. our eyes are light detectors.

Intensity Decreases as Distance from the Source Increases

1) When radiation is absorbed by matter, the photons transfer their energy to the matter.
2) The energy 'deposited' by a beam of photons depends on how many photons there are and the energy of each photon. (Total energy = number of photons × energy of each photon.)
3) The intensity of radiation hitting a surface means how much energy arrives at each square metre of surface per second.
4) The units of intensity are W/m^2 — watts per square metre.
5) The intensity of a beam of radiation decreases with distance from the source.
6) For example, when you stand near a fire you feel nice and warm. If you moved further away, you'd feel a lot colder — because fewer photons would be reaching you.

Intensity decreases with distance from ketchup...

Remember, different types of EM radiation behave differently. E.g. radio is transmitted through brick walls but light isn't. The other thing to remember from this page is the stuff about intensity. As you get further from the source, intensity decreases. That's why it's cold on Pluto — it's so far from the Sun.

Ionisation

Generally, high-energy EM radiation is more harmful than low-energy radiation. Here's why.

Some EM Radiation Causes Ionisation

1) All substances are made of atoms (see p.83). When a photon hits an atom or molecule, it sometimes transfers enough energy to break the molecule (or atom) into bits called ions.

2) This process is called ionisation:

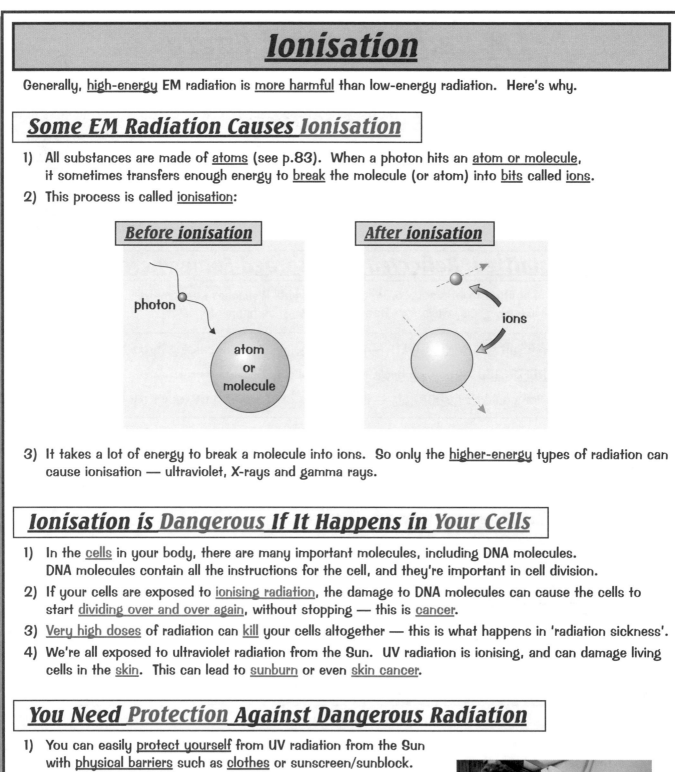

Before ionisation

photon

atom or molecule

After ionisation

ions

3) It takes a lot of energy to break a molecule into ions. So only the higher-energy types of radiation can cause ionisation — ultraviolet, X-rays and gamma rays.

Ionisation is Dangerous If It Happens in Your Cells

1) In the cells in your body, there are many important molecules, including DNA molecules. DNA molecules contain all the instructions for the cell, and they're important in cell division.

2) If your cells are exposed to ionising radiation, the damage to DNA molecules can cause the cells to start dividing over and over again, without stopping — this is cancer.

3) Very high doses of radiation can kill your cells altogether — this is what happens in 'radiation sickness'.

4) We're all exposed to ultraviolet radiation from the Sun. UV radiation is ionising, and can damage living cells in the skin. This can lead to sunburn or even skin cancer.

You Need Protection Against Dangerous Radiation

1) You can easily protect yourself from UV radiation from the Sun with physical barriers such as clothes or sunscreen/sunblock.

2) The ozone layer — a layer of gas in the atmosphere — also protects us from UV radiation. It absorbs some of the UV high up in the atmosphere, so it never reaches us here on the surface.

3) When you have an X-ray taken, the radiographer might put lead shields over parts of your body that aren't being investigated. The lead absorbs X-rays, protecting you from unnecessary exposure. Radiographers also wear lead aprons to protect themselves.

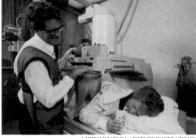

LARRY MULVEHILL / SCIENCE PHOTO LIBRARY

Use protection — wear a hat...

There's no point being paranoid about the dangers of radiation. Being exposed to high doses of X-rays doesn't mean you'll definitely get cancer — it increases the risk, but it's well worth having an X-ray to find out if you're seriously injured or ill. Sunbathing for hours with no protection is just stupid, though.

Some Uses of EM Radiation

No phones, no dinner — what would we do without EM radiation?

EM Radiation Can Cause Heating

1) Non-ionising radiation, e.g. light, doesn't have enough energy to break atoms up. When it's absorbed by a substance it transfers energy to the atoms or molecules of the substance — and heats them up.

2) The more intense the radiation (see p.55), the greater the heating effect.

3) This heating effect can damage living cells, e.g. you get burned if you absorb too much infrared radiation.

4) Heating can be pretty useful though — it's how we cook food, after all. 'Normal' ovens do this with infrared radiation and microwave ovens do it with microwaves (surprise):

Microwaves heat up anything which contains particles that the microwaves can make vibrate. Some microwaves can heat water molecules. This is handy, because there's water in all food substances.

Microwave ovens come in different power ratings (e.g. 800 W, 1200 W). More powerful ovens produce higher intensity radiation, so less time is needed to produce the same heating effect — food cooks more quickly.

Microwave radiation would heat up the water in your body's cells if you were exposed to it. Microwave ovens have metal cases and screens over their glass doors which stop the microwaves getting out.

EM Radiation Can Transmit Information

EM radiation has been used to send information for years — e.g. using light to send signals in Morse code. There are quite a few more modern uses as well:

Infrared	TV remote controls 'Night vision' cameras
Microwave	Mobile phones, Satellite communication
Radio	TV and Radio transmissions Radar

Some People Say There are Health Risks with Using Microwaves

1) When you make a call on your mobile, the phone emits microwave radiation. Some of this radiation is absorbed by your body, and causes heating of your body tissues (which all contain water).

2) There are concerns that heating of tissues like the brain and jaw could increase the risk of some medical conditions, possibly including cancers. There is no conclusive evidence to show this, though.

3) The radiation is quite low intensity, so the heating is probably very minor and nothing to be too worried about.

4) Mobile phone masts also emit microwave radiation. Some people who live very close to masts are worried about their possible effects. Again, there isn't much evidence — and any health problems might take a long time to emerge, so we might not know either way for many years.

Microwaves — for when you're only slightly sad to say goodbye...

Ovens and mobile phone networks both use microwaves, but with different energies. In an oven, the whole idea is for water molecules to absorb the microwave energy. But that's not ideal with a phone network — if the energy were absorbed by water molecules, the signals would never get through clouds.

Module P2 — Radiation and Life

EM Radiation and Life

No sunlight = no life. (Apart from a few very odd things right at the bottom of the sea, that is.)

Some Radiation from the Sun Passes Through the Atmosphere

1) The Earth is surrounded by an atmosphere made up of various gases — the air.

2) As you go further from the Earth's surface, the air gets thinner. (Mountaineers need to breathe oxygen from cylinders at very high altitudes because there isn't enough oxygen in the air to keep them alive.)

3) The gases in the atmosphere filter out certain types of radiation from the Sun — they absorb or reflect the radiation, so it never reaches the Earth's surface.

4) However, some types of radiation — mainly visible light and some radio waves — pass through the atmosphere quite easily.

The atmosphere gets thinner and thinner until there's only space.

Some radiation passes through the atmosphere...

...but some doesn't.

Radiation Makes Photosynthesis (and Most Life) Possible

Almost all life on Earth depends on EM radiation from the Sun reaching us. This radiation does two jobs:

1) It helps keep the planet warm — during the day, the Earth's surface absorbs radiation and warms up.

2) Sunlight provides the energy for photosynthesis (the process by which green plants make their food — see below).

3) So, without the Sun, the planet would be too cold for life as we know it. There'd also be no photosynthesis — so plants wouldn't grow, so animals couldn't feed on plants, so there'd be nothing to eat for anyone.

Photosynthesis Adds Oxygen to the Atmosphere

Photosynthesis is the process by which green plants make their food. This is what happens:

1) The plant takes in water from the soil and carbon dioxide from the air.

2) The plant's leaves (and other green parts, like the stem) absorb sunlight — so the energy of the light is transferred to the plant.

3) This energy is used in a chemical reaction between the water and the carbon dioxide.

4) The products of the reaction are sugar and oxygen. (Sugar is the plant's source of food.)

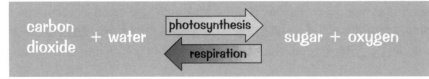

carbon dioxide + water → photosynthesis / ← respiration → sugar + oxygen

5) So when plants photosynthesise, they remove carbon dioxide from the atmosphere and add oxygen.

6) This is the reverse of respiration (a reaction between glucose and oxygen that releases carbon dioxide and water into the atmosphere) which goes on in all living things.

Let there be light...

The Sun emits all types of radiation, from radio to gamma rays. Most of the photons emitted in our direction never reach us though — they're absorbed by the atmosphere. The amount of light reaching us varies depending on how cloudy it is, but there's always enough to keep photosynthesis ticking over.

Module P2 — Radiation and Life

The Greenhouse Effect

The atmosphere keeps us warm by trapping heat.

The Greenhouse Effect Helps Regulate Earth's Temperature

1) The Earth absorbs EM radiation from the Sun (see previous page). This warms the Earth's surface up. The Earth then emits some of this EM radiation back out into space — this tends to cools us down.

2) Most of the radiation emitted from Earth is infrared radiation — heat.

3) A lot of this infrared radiation is absorbed by atmospheric gases, including carbon dioxide.

4) These gases then re-radiate heat in all directions, including back towards the Earth.

5) So the atmosphere acts as an insulating layer, stopping the Earth losing all its heat at night.

6) This is known as the 'greenhouse effect'. (In a greenhouse, the sun shines in and the glass helps keep some of the heat in.)
Without the greenhouse gases (e.g. CO_2) in our atmosphere, the Earth would be a lot colder than it is.

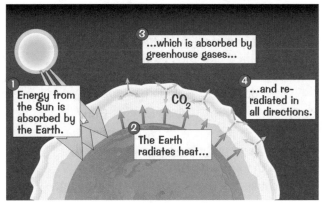

① Energy from the Sun is absorbed by the Earth.

② The Earth radiates heat...

③ ...which is absorbed by greenhouse gases...

④ ...and re-radiated in all directions.

Upsetting the Greenhouse Effect Could Lead to Climate Change

1) Since we started burning fossil fuels in a big way, levels of carbon dioxide in the atmosphere have risen drastically (see next page).

2) The 'greenhouse' is now working too well and we're starting to overheat.

3) There's a lot of evidence to show that global temperatures have risen over the last century or so — and even a small temperature rise could have a big effect on the world's climate.

The Consequences of Global Warming Could be Pretty Serious

Global warming could have some nasty consequences. But there's a lot of uncertainty about what's most likely to happen and when and how serious the consequences would be:

1) As the sea gets warmer, it will expand, causing sea levels to rise. This would be bad news for people living in low-lying places, like the Netherlands, East Anglia and the Maldives — they could be flooded.

2) Higher temperatures make ice melt. Water that's currently 'trapped' on land (as ice) could run into the sea, causing sea levels to rise even more.

3) Weather patterns might change in many parts of the world. It's thought that many regions will suffer more extreme weather, e.g. longer, hotter droughts.

4) Hurricanes form over water that's warmer than 26 °C — so with more warm water, you'd expect more hurricanes.

5) These kinds of changes will affect food production — some regions would become too dry to grow food, some too wet.

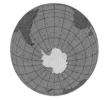

There's a lot of ice at the poles now, but is it melting?

Climate control — yeah, it's standard in most 4×4s...

'Global warming' could mean that some parts of the world cool down. For instance, as ice melts, lots of cold fresh water will enter the sea and this could disrupt the ocean currents. This would be bad news for us in Britain — if the nice warm currents we get at the moment weaken, we'll be a lot colder.

The Carbon Cycle

Global warming and carbon emissions — you can't avoid hearing about them, so read on.

The Carbon Cycle Shows How Carbon is Naturally Recycled

Two of the 'greenhouse gases' which keep the Earth warm are carbon dioxide (CO_2) and methane (CH_4). There's a fairly small amount of CO_2 and just trace amounts (a tiny tiny bit) of methane. Both of these gases contain carbon.

1) All the carbon on Earth moves in a big cycle.

2) All plants and animals contain carbon. When they die, they start to decay — bacteria and fungi (decomposers) break them down into various compounds.

3) During decomposition, oxygen from the air combines with carbon from the plant, and carbon dioxide is released into the atmosphere. (This happens because the decomposers are respiring.)

4) Other processes also release carbon dioxide into the air — including respiration in living plants and animals and burning (p.17-19).

5) Photosynthesis does the opposite — it removes carbon dioxide from the atmosphere (see p.58).

6) For thousands of years, these processes have all balanced out — and carbon dioxide has been removed from the air and added to the air in approximately equal quantities.

7) So the concentration of CO_2 in the atmosphere has been about the same for thousands of years. But recently that's been changing...

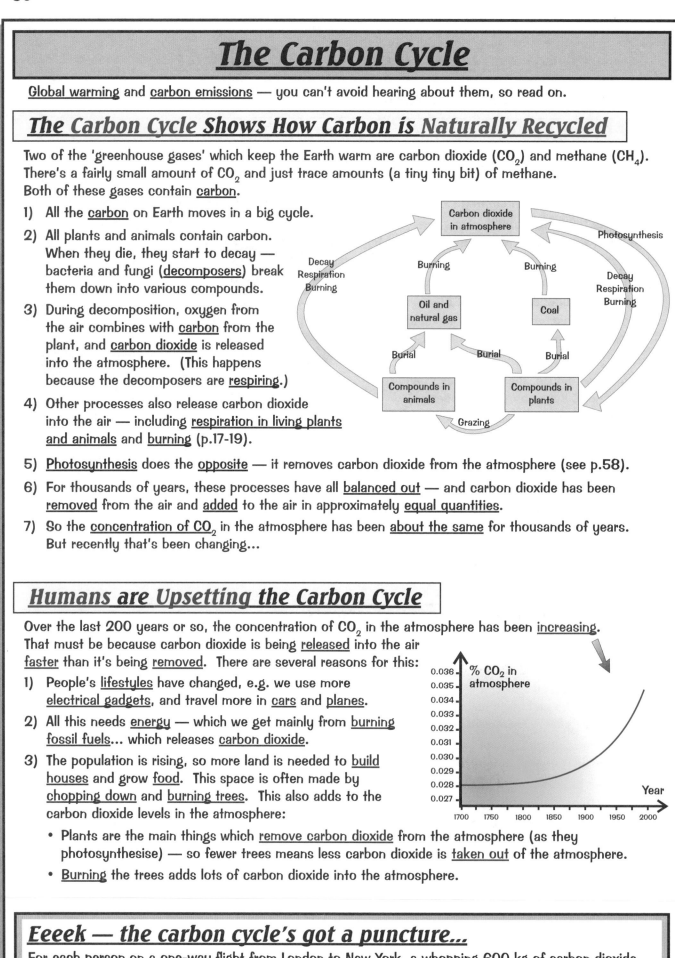

Humans are Upsetting the Carbon Cycle

Over the last 200 years or so, the concentration of CO_2 in the atmosphere has been increasing. That must be because carbon dioxide is being released into the air faster than it's being removed. There are several reasons for this:

1) People's lifestyles have changed, e.g. we use more electrical gadgets, and travel more in cars and planes.

2) All this needs energy — which we get mainly from burning fossil fuels... which releases carbon dioxide.

3) The population is rising, so more land is needed to build houses and grow food. This space is often made by chopping down and burning trees. This also adds to the carbon dioxide levels in the atmosphere:

- Plants are the main things which remove carbon dioxide from the atmosphere (as they photosynthesise) — so fewer trees means less carbon dioxide is taken out of the atmosphere.

- Burning the trees adds lots of carbon dioxide into the atmosphere.

Eeeek — the carbon cycle's got a puncture...

For each person on a one-way flight from London to New York, a whopping 600 kg of carbon dioxide is added to the air. You can now pay to plant some trees to try and 'soak up' the carbon emissions you're responsible for. This sounds great, but there can be problems, e.g. for people living where the new plantations are planned. It might be better to release less CO_2 from fossil fuels in the first place.

EM Radiation and Risk

There are some risks that an individual person can't do much about — like climate change, for example. With other dangers, like harmful UV radiation from the Sun, you can minimise the risks quite easily.

Human Activities are Probably Causing Global Warming

1) There is a correlation (link) between the concentration of carbon dioxide in the atmosphere and global temperature. When carbon dioxide levels are higher, temperatures are also higher.

2) Over the last 150 years or so, global temperatures have been rising ('global warming') and carbon dioxide levels have also been rising.

3) We don't know for sure that one is causing the other, but the evidence seems to point that way.

4) If we assume that an increase in carbon dioxide levels is causing global warming, we can try to reduce the risk of climate change by:

> Burning less fossil fuels, e.g.
> - using less energy in homes and businesses (e.g. using energy-efficient appliances, insulating buildings so they need less heating in winter, switching off appliances when not in use).
> - using wind, solar and other renewable resources to generate electricity, rather than burning fossil fuels (see page 89).
>
> Taking more carbon dioxide out of the atmosphere, e.g.
> - replacing trees that have been cut down for wood or to make room for farming.

Sunbathing is a Risky Business Too

1) There's strong evidence to show that prolonged exposure to UV radiation (in sunlight) causes skin cancer.

2) This doesn't mean that everyone who spends a certain amount of time in the sun will get skin cancer. Your chances of developing skin cancer depend on many different factors — e.g. skin type, how often you got sunburnt as a child, use of sunscreen. We know this from various studies:

> E.g. A scientist might want to see whether people with very freckly skin are at higher risk of skin cancer. The study would have to compare the rates of skin cancer in two groups — people with lots of freckles and people with few or no freckles. The study would have to be designed carefully.
>
> 1) The samples have to be big enough — several thousand people rather than ten or twenty.
>
> 2) The sample groups must be well matched so that other factors don't influence the results. E.g. older people are more likely to develop skin cancer — so both samples should have the same number of people in each age group.

3) But some people sunbathe for hours on end, without using sunscreen — even though they know it increases their chances of getting cancer. It seems they're prepared to take the risk. That might be because they feel there are benefits (like getting a fashionable tan) which outweigh the risks.

Life — it's a risky business...

Finding correlations is an important part of many scientists' work. Remember, one piece of evidence isn't enough to show a link. For instance, noticing that 'I had cornflakes for breakfast' and 'I was late to school today' doesn't mean there's a correlation between cornflake-eating and lateness to school.

Revision Summary for Module P2

So, electromagnetic radiation — it's a bit weird, but if you can answer all these questions you're well on the way to a happy and fulfilled relationship with module P2.

1) What does radiation transfer?

2) Write out the electromagnetic spectrum, starting with radio waves.

3) What is a photon?

4) Which has more energy — an infrared photon or an ultraviolet photon?

5) Name three sources of electromagnetic radiation.

6) State three things that might happen to radiation when it meets some matter.

7) When radiation is absorbed, what happens to its energy?

8) What does 'intensity' mean?

9) Describe how the intensity of radiation changes with distance from the source.

10) What is ionisation?

11) Name the three types of electromagnetic radiation that can cause ionisation.

12) Describe how ionising radiation can damage living cells.

13) Explain why excessive sunbathing can increase the risk of skin cancer.

14) Outline the measures people can take to minimise their exposure to radiation from:
    a) the Sun, b) hospital X-rays.

15) Explain why the ozone layer is important for life on Earth.

16) Explain why microwaves aren't ionising.

17) What effect does non-ionising radiation have on a substance that absorbs it?

18) Non-ionising radiation can damage living cells. Explain how.

19) Describe how a microwave oven cooks food.

20) Why do microwave ovens have door screens?

21) Apart from microwave ovens, give another use of microwaves.

22) State two uses of: a) infrared radiation, b) radio waves.

23)*Some people think that using a mobile phone for long periods of time might pose a health risk.
    Explain why.

24) The Earth's atmosphere stops a lot of radiation from reaching the surface of the planet.
    Describe how this happens.

25) Why is radiation from the Sun so important for life on Earth?

26) Outline the process of photosynthesis.

27) What are the products of respiration?

28) Name one 'greenhouse gas'.

29) Explain why we need some greenhouse gases in the atmosphere.

30) Explain how higher levels of carbon dioxide in the atmosphere are thought to lead to global warming.

31) Explain why global warming is expected to cause a rise in sea levels.

32) Why might rising sea levels be a problem?

33) Describe two other major problems that might be caused by climate change.

34) Name two processes which add carbon dioxide to the atmosphere.

35) Suggest two reasons why there is more carbon dioxide in the atmosphere now than 200 years ago.

* Answer on page 100.

Evolution

So, how did we get from primordial soup to poodles? The glories of evolution...

Life on Earth is Incredibly Varied

1) There's an <u>enormous</u> number of <u>species</u> on Earth.

2) Scientists don't know <u>how many</u> species there are
   — estimates vary from <u>2 million</u> to <u>100 million</u>.

3) There are also loads of species that have become <u>extinct</u>, so since life
   began, a huge number of different living things have existed on the planet.

4) The very <u>first</u> living things were very <u>simple</u>. Life then evolved to become more <u>complex</u> and <u>varied</u>.

5) All living things that exist now (and species that are now extinct) <u>evolved</u> from those
   very simple <u>early life forms</u>.

Algae are good examples of early, simple life forms.

There is Good Evidence for Evolution

If you're going to say that living things <u>evolved</u> from very simple life forms, you need
to find good <u>evidence</u> for it. <u>Fossil records</u> and <u>DNA</u> both provide <u>evidence</u> for evolution:

There's evidence for evolution in the <u>fossil record</u>, which shows
species getting more and more <u>complex</u> as time goes on.

<u>DNA</u> controls the <u>characteristics</u> of living things.
All living things have some <u>similarities</u> in their DNA, as
you would expect if they have all <u>evolved</u> from the <u>same</u>
simple life forms.

The more <u>closely related</u> two species are, the more
<u>similar</u> their DNA is. Scientists can use the <u>similarities</u>
<u>and differences</u> in DNA to work out how life has <u>evolved</u>.

<u>Humans</u> and <u>chimpanzees</u> are closely related, and about <u>95%</u> of their DNA is the same.

Somehow, a Long Time Ago, Life Must Have Started

Scientists estimate that <u>life</u> on Earth <u>began</u> about <u>3500 million years ago</u>.
There are three main ideas about <u>how</u> life first appeared on Earth:

1) Many different <u>religious groups</u> believe that life was created by a <u>god</u>.
   The existence of God <u>cannot</u> be proved or disproved.

2) Some people believe that simple life forms arrived on Earth from <u>space</u>, possibly in a <u>meteorite</u>.

3) Most scientists believe that life began when a <u>chemical</u> or group of chemicals was
   formed that could <u>copy itself</u> and so '<u>reproduce</u>'. Exactly how this could've happened
   is still uncertain. But, experiments have shown that if you expose a mixture of
   <u>chemicals</u> that were around on ancient Earth (water, ammonia, methane and hydrogen)
   to an <u>electrical discharge</u> (like <u>lightning</u>), <u>amino acids</u> can form. Amino acids are the
   building blocks of <u>proteins</u>, which are essential to <u>every living thing</u> on Earth.

It's life, Jim, but it's a bit hairy for my liking...

If you <u>know</u> the man in the picture, don't point and laugh at him. I thought it was funny to label him
'chimp', but, well, he might not find it all that funny himself. There were loads of <u>famous</u> people I
fancied putting in the book and labelling 'chimp', but sadly I'm not allowed to for legal reasons... Hmmf.

Natural Selection

This is important — read it carefully. <u>Natural selection</u> is an important <u>process</u> that causes <u>evolution</u>.

Natural Selection Means the "Survival of the Fittest"

It works like this:

1) Living things show <u>variation</u> — they're <u>not</u> all the same. OK, it's fairly simple so far.

2) Some <u>organisms</u> will be <u>better suited</u> to their environment (e.g. they can evade predators better or outcompete other things for food) and so will have a <u>better chance</u> of survival.

3) Those organisms will then have an increased chance of <u>breeding</u> and passing on their <u>genes</u>. This means that a <u>greater</u> proportion of individuals in the next generation are likely to have the better <u>alleles</u> that help <u>survival</u>.

4) Over many generations, the species becomes better and better able to <u>survive</u>. The 'best' features are <u>naturally selected</u> and the species becomes better and better <u>adapted</u> to its environment.

| HERE'S AN EXAMPLE |

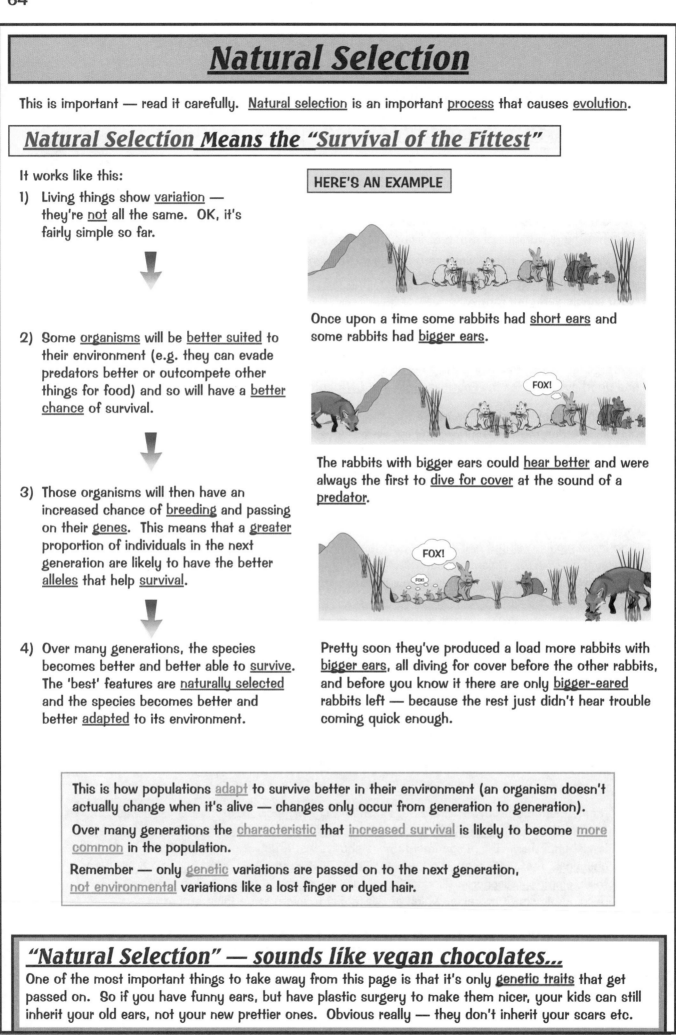

Once upon a time some rabbits had <u>short ears</u> and some rabbits had <u>bigger ears</u>.

FOX!

The rabbits with bigger ears could <u>hear better</u> and were always the first to <u>dive for cover</u> at the sound of a <u>predator</u>.

FOX!

FOX!

Pretty soon they've produced a load more rabbits with <u>bigger ears</u>, all diving for cover before the other rabbits, and before you know it there are only <u>bigger-eared</u> rabbits left — because the rest just didn't hear trouble coming quick enough.

This is how populations <u>adapt</u> to survive better in their environment (an organism doesn't actually change when it's alive — changes only occur from generation to generation).

Over many generations the <u>characteristic</u> that <u>increased survival</u> is likely to become <u>more common</u> in the population.

Remember — only <u>genetic</u> variations are passed on to the next generation, <u>not environmental</u> variations like a lost finger or dyed hair.

"Natural Selection" — sounds like vegan chocolates...

One of the most important things to take away from this page is that it's only <u>genetic traits</u> that get passed on. So if you have funny ears, but have plastic surgery to make them nicer, your kids can still inherit your old ears, not your new prettier ones. Obvious really — they don't inherit your scars etc.

Natural Selection

So thanks to natural selection organisms are able to evolve in response to changes in the environment. But just what would have happened if everything hadn't evolved in the way it did. We might have been walking around with six arms and purple and green skin — cool.

Life on Earth Could Easily Have been Very Different

The different species that exist on Earth today are the result of the changing conditions that the planet has gone through and how organisms have adapted to those changes. If some of the events in Earth's history had been different, then different types of organism would have been produced.

EXAMPLE

The dinosaurs became extinct about 65 million years ago due to some sort of environmental change — a meteor impact, climate change or maybe a combination of factors. Dinosaurs were unable to evolve at a quick enough rate to cope with this environmental change and so became extinct. If this hadn't happened, the species that we'd see on Earth today would be very different from the ones we now have.

Selective Breeding is Where Humans Choose What Gets Selected

Selective breeding is when humans select the plants or animals that are going to breed and flourish, according to what we want from them. Here's how we do it:

1) Animals or plants with a desired feature, e.g. plants with high crop yield, are selected.
2) They're bred with each other.
3) The best of the offspring are selected and bred.
4) The process is repeated to develop the desired trait.

One difference between selective breeding and natural selection is that natural selection selects features that help survival. In selective breeding humans choose the features they want, which may not necessarily help the plant or animal's survival, e.g.

Increased milk production

A breeder might choose to only breed from the cows that produce the most milk, so that future generations of cows produce more milk than previous generations. This has no real survival advantage to the cow — but is advantageous to the farmer.

Increased number of offspring

Farmers can selectively breed sheep to increase the number of lambs born. Female sheep who produce large numbers of offspring are bred with male sheep whose mothers had large numbers of offspring.

Selective beading — making necklaces look pretty...

Part of me wishes dinosaurs hadn't become extinct — think how awesome it would be to have your own pet dino, riding it around, crushing stuff, eating people you don't really like. Anyway, enough of that — make sure you know the difference between natural selection and selective breeding.

A Scientific Controversy

This page has got an <u>exciting</u> title — 'a scientific controversy' — ooh, I love a bit of controversy I do.

Natural Selection was a Revolutionary Idea...

<u>Charles Darwin</u> (1809–1882) was the first to suggest the theory of <u>natural selection</u>. He came up with it through a combination of <u>imagination</u>, <u>opportunity</u> and <u>creativity</u>:

1) He had the <u>imagination</u> to see <u>beyond</u> the idea that all species are <u>unchangeable</u>, which is what nearly all scientists believed at the time.

2) His visit to the <u>Galapagos islands</u> put him in an area where there were lots of <u>islands</u>, close together yet <u>isolated</u> by the sea — <u>ideal conditions</u> for the evolution of <u>new species</u>.

3) Given this opportunity, he made careful <u>observations</u> and then applied a great deal of <u>analytical</u> and <u>creative thinking</u> to then come up with the idea of <u>natural selection</u>.

...Which Upset Quite a Few People

<u>Religious</u> people (i.e. most people at the time) <u>disliked</u> Darwin's ideas because his ideas disagreed with their belief that all species were individually created by <u>God</u>. However, that wasn't the only <u>reason</u> for his theory being rejected at first. There are <u>two</u> main reasons why many scientists <u>disagreed</u> with Darwin at first:

1) Natural selection is a <u>theory</u>. It's based on the <u>interpretation</u> of observations. Different people can interpret things <u>differently</u> and there's always likely to be <u>disagreement</u> when there's no <u>absolute proof</u>.

2) When new data <u>challenges</u> an accepted idea, you have to assume that <u>either</u> the <u>data</u> is wrong or the <u>original idea</u> is wrong. It would have been hard to change people's minds about where different species came from, so many would have automatically <u>assumed</u> that the <u>data</u> was wrong.

Not Everyone Thinks the Theory of Natural Selection is True

Many scientists are <u>convinced</u> that natural selection is the best explanation for evolution, but some <u>aren't sure</u>, and some people (e.g. <u>creationists</u>) don't believe that <u>evolution</u> has happened at all. It's hard for some people to make their minds up about natural selection. Here's why:

1) The <u>data</u> may be <u>reliable</u>, but most reports only offer one <u>interpretation</u> of the data, when others may be possible. Scientists may also choose to <u>ignore</u> any data that <u>conflicts</u> with their own ideas.

2) Natural selection seems a <u>good</u> explanation for the way things <u>adapt</u> to their environment — it can explain how many features of animals and plants could have appeared. However, it's <u>not so good</u> at explaining the evolution of <u>some features</u>, or (necessarily) the <u>rate</u> of evolution. (Opponents concentrate on these <u>weaker</u> areas to challenge the <u>whole</u> idea of natural selection.)

3) Ideas about exactly how natural selection works have <u>changed</u> over the years and are <u>still changing</u> as <u>new data</u> becomes available.

Not as controversial as I was expecting...

I feel a bit sorry for Darwin. He came up with such a great <u>theory</u>, but didn't publish it for ages. Apparently, he knew the reaction he'd get from the religious community, so he was planning on leaving the publication of his theory until after his death. It was <u>only</u> when some other scientist came up with a similar theory that Darwin thought it was time to get on and publish. And it all kicked off...

Human Evolution

Well I don't know if it's just me, but I think this section is really interesting. It's amazing to think that your super-distant relatives were apes — before that, we were squirrelly things, and before that, fish... Er, maybe.

The Big Human Brain Gave a Big Survival Advantage

We human beings have a brain that is very large in relation to the rest of our body, compared to other animals. The evolution of this larger brain gave ancient humans a big advantage when it came to survival:

1) It allowed them to solve problems better, and to develop effective tools. This would have been particularly useful in hunting, allowing them to capture more food.

2) The larger brain also allowed humans to develop better ways of communicating, so that ideas could be shared. This would have been useful in group hunting.

The Human 'Family Tree' Shows Where We Came From

The diagram shows a recent idea on how humans (Homo sapiens) evolved.

You don't need to remember all the details of the diagram, but the following basic principles are important:

1) There have been quite a lot of different human-like species (hominids) on the planet, and all except Homo sapiens (us) are now extinct.

2) All hominids are thought to have evolved from a common ancestor, which lived a few million years ago.

3) Hominids have evolved in different ways, as shown by the branching in the diagram.

```
        Australopithecus
           afarensis
               |
          Homo erectus
               |
        ┌──────┴──────┐
   Homo erectus      Homo
   (Java man)    heidelbergensis
                      |
               ┌──────┴──────┐
            Homo          Homo sapiens
         neanderthalensis
```

The Theory of Human Evolution Might be Wrong

The problem with working out how humans evolved is that all the other human species are extinct, and the only evidence we have about them is often just a few bones.

The human family tree has been worked out on the basis of similarities and differences between bones found in different places (among other things). Here's an example which proves you shouldn't believe every piece of evidence found:

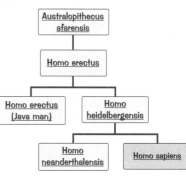

More evidence like this would be useful.

PILTDOWN MAN

In 1912, the discovery of skull and jawbone fragments seemed to provide evidence of a missing link between apes and humans. This evidence became known as 'Piltdown man' after the English village where it was found. The jawbone was ape-like, but the skull was like that of a modern man. The evidence suggested that humans might have evolved in a different way from what is now believed. However, about 30 years later it was found to be a forgery. Shocking news. The Piltdown man evidence didn't fit in with the theory of human evolution anyway — it was the observation and not the theory that was proved to be incorrect.

I know some teachers who look like they need to evolve...

You don't need to remember all the details about Piltdown man. He's just there as an example to show you that if an observation seems to change an existing theory, then either the observation or the theory must be wrong. In this case, Piltdown man was a fake — I wonder who made him and why they did it.

Module B3 — Life On Earth

Communication

We were once really <u>simple</u> organisms, made of just a few cells, and now we have <u>eyes</u> and <u>lungs</u> and a massive great <u>brain</u> capable of so many exciting things. It makes you wonder at the power of nature.

Evolution Eventually Led to Some Very Complicated Organisms

As multicellular organisms got <u>bigger</u>, they got more <u>complicated</u>, developing different parts that were <u>specialised</u> for different jobs. Once that happened, they needed ways to <u>communicate</u> with and <u>coordinate</u> the different parts. To do this, animals evolved <u>nervous</u> and <u>hormonal</u> communication systems. The nervous system is for <u>fast</u>, <u>short-lived</u> responses like moving muscles.

The Nervous System Lets You React to Stimuli

A <u>stimulus</u> is a <u>change in your environment</u> which you may need to react to (e.g. an approaching tiger). You need to be constantly monitoring what's going on so you can respond if you need to.

1) Detecting stimuli is the job of your <u>sense organs</u>.

2) You have five different sense organs — <u>eyes</u>, <u>ears</u>, <u>nose</u>, <u>tongue</u> and <u>skin</u>.

3) They all contain different <u>receptors</u>. Receptors are groups of cells which are <u>sensitive</u> to a <u>stimulus</u>. They change <u>stimulus energy</u> (e.g. light energy) into <u>electrical impulses</u>.

4) A stimulus can be <u>light</u>, <u>sound</u>, <u>pressure</u>, a <u>chemical</u>, <u>temperature</u> or a change in <u>position</u>.

> There's more about how all these things work together on the next page.

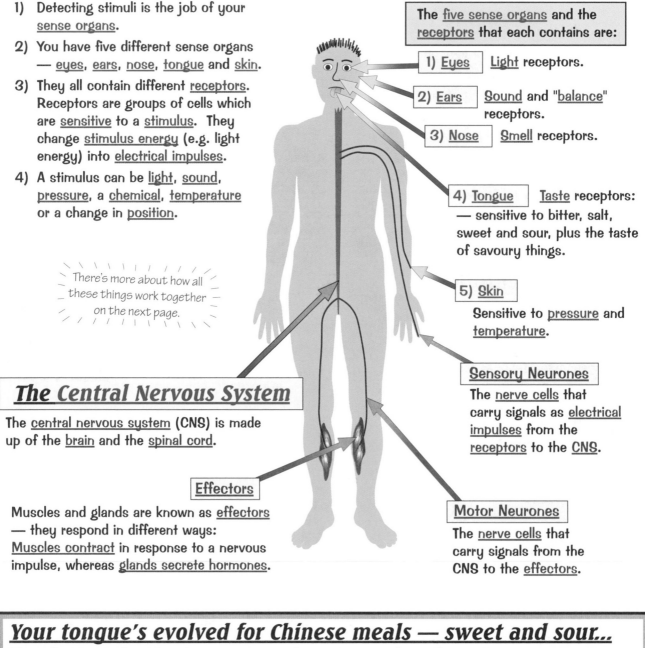

The <u>five sense organs</u> and the <u>receptors</u> that each contains are:

1) <u>Eyes</u> — <u>Light</u> receptors.

2) <u>Ears</u> — <u>Sound</u> and "<u>balance</u>" receptors.

3) <u>Nose</u> — <u>Smell</u> receptors.

4) <u>Tongue</u> — <u>Taste</u> receptors: — sensitive to bitter, salt, sweet and sour, plus the taste of savoury things.

5) <u>Skin</u> — Sensitive to <u>pressure</u> and <u>temperature</u>.

Sensory Neurones

The <u>nerve cells</u> that carry signals as <u>electrical impulses</u> from the <u>receptors</u> to the <u>CNS</u>.

The Central Nervous System

The <u>central nervous system</u> (CNS) is made up of the <u>brain</u> and the <u>spinal cord</u>.

Effectors

Muscles and glands are known as <u>effectors</u> — they respond in different ways: <u>Muscles contract</u> in response to a nervous impulse, whereas <u>glands secrete hormones</u>.

Motor Neurones

The <u>nerve cells</u> that carry signals from the CNS to the <u>effectors</u>.

Your tongue's evolved for Chinese meals — sweet and sour...

It's really important not to get <u>sense organs</u> and <u>receptors</u> mixed up: The <u>eye</u> is a <u>sense organ</u> — it contains <u>light receptors</u>. The <u>ear</u> is a <u>sense organ</u> — it contains <u>sound receptors</u>.

Communication

Nervous impulses are great, but they're not the only way messages are sent around your body. There are also chemical messengers called <u>hormones</u>.

The Central Nervous System Coordinates Information

OK, this is what happens when you <u>detect a change</u> in your environment (a <u>stimulus</u>):

1) Your <u>sensory neurones</u> carry the information <u>from receptors</u> (e.g. light receptors in the eye) to the <u>CNS</u>.

2) The CNS then <u>figures out what to do</u> about the information. It's basically the control centre for your body and it <u>coordinates</u> all the information you receive from your sense organs.

3) The CNS sends information about how to respond to an <u>effector</u> (muscle or gland) along a <u>motor neurone</u>.

4) The effector <u>responds</u> accordingly.

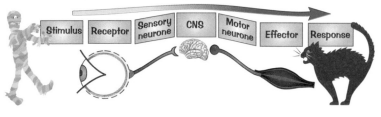

Stimulus → Receptor → Sensory neurone → CNS → Motor neurone → Effector → Response

Hormones <u>are</u> Chemical Messengers <u>Sent in the</u> Blood

1) <u>Hormones</u> are <u>chemicals</u> released directly into the <u>blood</u>. They are carried in the <u>blood plasma</u> to other parts of the body, but only affect particular cells (called <u>target cells</u>) in particular places.

2) Hormones control things in organs and cells that need <u>constant adjustment</u>.

3) They are produced in various <u>glands</u> and travel through your body at "<u>the speed of blood</u>".

4) Hormones tend to have relatively <u>long-lasting</u> effects (compared to <u>nerves</u> whose effects don't last long at all).

Learn These Examples of <u>Communication</u> in your Body

Examples of <u>nervous communication</u> are:

1) <u>Picking something up</u> — your <u>brain</u> instructs the <u>muscles</u> of your arm, hand and fingers to move in the desired way.

2) <u>Vision</u> — your <u>eyes</u> detect patterns of light and then send signals to the <u>brain</u>, which builds up a picture of what you are seeing.

Examples of <u>hormonal communication</u> are:

1) When you are <u>scared or angry</u>, your <u>adrenal glands</u> produce the hormone <u>adrenaline</u>. This gets your body <u>revved up</u>, ready for action — it increases your <u>breathing</u> and <u>heart rate</u>, and directs blood to your <u>muscles</u>.

2) When your <u>blood sugar</u> rises after a meal, the <u>pancreas</u> releases the hormone <u>insulin</u>. This acts all over your body to help you <u>use</u> or <u>store</u> the sugar.

Who knew being scared was so organised...

Make sure you know the differences between <u>nervous</u> and <u>hormonal</u> communication systems. The nervous system uses <u>electrical impulses</u> for <u>fast</u>, <u>short-lived</u> responses. Hormones are a lot <u>slower</u> — they travel in the <u>blood</u> and the responses they bring about are generally <u>longer-lasting</u>. Easy peasy.

Interdependence

As a slight detour from nerves and hormones, here's some stuff on <u>interdependence</u>.
Interdependence is about how nearly all species are <u>dependent</u> on a number of other species for <u>survival</u>.

Every Living Thing Needs Resources from Its Environment

The <u>environment</u> in which an organism lives provides it with <u>factors</u> that are <u>essential</u> for life. These include:

If any essential factor in a habitat is in <u>short supply</u>, the different species that need it have to <u>compete</u> for it. If there's not enough to go around, some organisms <u>won't survive</u>. This will <u>limit</u> the size of their populations in that habitat.

1) <u>Light</u> (needed by plants to make food)
2) <u>Food</u> (for animals) and <u>minerals</u> (for plants)
3) <u>Oxygen</u> (for animals and plants) and <u>carbon dioxide</u> (for plants)
4) <u>Water</u> (vital for all living organisms)

Organisms also depend on <u>other organisms</u> (usually for food) — this is called <u>interdependence</u>.

Any Change in Any Environment Can Have Knock-on Effects...

The <u>interdependence</u> of all the living things in a habitat means that any major <u>change</u> in the habitat can have <u>far-reaching effects</u>.

The diagram on the right shows part of a <u>food web</u> (a diagram of what eats what) from a <u>stream</u>.

<u>Stonefly larvae</u> are particularly sensitive to <u>pollution</u>. Suppose pollution <u>killed</u> them in this stream. The table below shows some of the <u>effects</u> this might have on some of the other organisms in the food web.

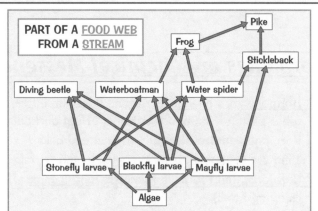

PART OF A <u>FOOD WEB</u> FROM A <u>STREAM</u>

Pike, Frog, Stickleback, Diving beetle, Waterboatman, Water spider, Stonefly larvae, Blackfly larvae, Mayfly larvae, Algae

Organism	Effect of loss of stonefly larvae	Effect on population
Blackfly larvae	Less competition for algae	Increase
	More likely to be eaten by predators	Decrease
Water spider	Less food	Decrease
Stickleback	Less food (if water spider or mayfly larvae numbers decrease)	Decrease

Remember that food webs are <u>very complex</u> and that these effects are difficult to predict accurately.

...Even Possibly Extinction

The fossil record contains many species that <u>don't exist any more</u> — these species are said to be <u>extinct</u>. <u>Dinosaurs</u> and <u>mammoths</u> are extinct, with only <u>fossils</u> and other remains to tell us they existed at all.

> <u>Rapid change</u> can cause a species to become <u>extinct</u>.
> Here are <u>three</u> changes that could cause the extinction of a species:
>
> 1) The <u>environmental conditions</u> change (e.g. destruction of habitat).
> 2) A new species is introduced which is a <u>competitor</u>, <u>disease organism</u> or <u>predator</u> of that species (this could include humans hunting them).
> 3) An organism in its <u>food web</u> that it is reliant on becomes <u>extinct</u>.

I'm dependent on the coffee plant...

If you're asked to analyse the <u>consequences</u> of a change in a food web, consider 'knock-on' effects as well as the organisms which are directly affected. You'll find that there are loads of different possible things that could happen — so just be aware of all the <u>possibilities</u> and you'll be sure of good marks.

Humans and the Earth

This page is about how we've caused <u>extinction</u> in the past, either <u>directly</u> by just blatantly killing things, or <u>indirectly</u> by doing other irresponsible things. It's also on why making species extinct is a <u>bad</u> idea.

Human Activity Can Directly Cause Extinction

Many organisms have become <u>extinct</u> in the past, and many are <u>threatened</u> with extinction now, often due to <u>human activities</u>. In some cases, humans have caused extinction <u>directly</u>. For example:

1) The <u>passenger pigeon</u> is thought to have once been the most <u>common</u> bird in the world, with an estimated five billion in North America. Numbers began to <u>decline</u> in the mid 1800s and the last bird died in captivity in 1914. Reasons for its <u>extinction</u> include <u>hunting</u> (both for sport and for food) and (to a lesser extent) <u>deforestation</u>.

2) The '<u>Tasmanian wolf</u>' was hunted to extinction on the island of Tasmania in the early 20th century. Humans first reduced its numbers by destroying its <u>habitat</u> to set up sheep <u>farms</u>. Then when the Tasmanian wolves started eating the sheep, the farmers <u>killed</u> the rest of them.

Human Activity Can Also Cause Extinction Indirectly

Humans can also cause extinction <u>indirectly</u> by destroying an organism's <u>habitat</u> or by introducing <u>new species</u> which it can't compete with. Here are a couple of examples:

(1) The <u>blue pike</u> from the Great Lakes in Canada and the US became extinct around 1970. Although it was hunted, the main reasons for its extinction were the <u>draining</u> and <u>pollution</u> of its wetland habitat and the introduction of <u>new species</u> which reduced blue pike numbers due to predation and competition.

(2) The <u>dusky seaside sparrow</u> became extinct in 1987. It lived in a restricted habitat in Florida in the USA. It only nested in <u>moist areas</u> and when one of its two major nesting sites was <u>flooded</u> and the other was <u>drained</u>, it became extinct.

Earth's Biodiversity is Important

<u>Biodiversity</u> is the number and variety of organisms found in an area. <u>Maintaining</u> biodiversity, rather than causing species to become extinct, is an important part of using the environment in a <u>sustainable</u> way. Here are three reasons why:

1) Every living organism plays a <u>role</u> in its ecosystem. Research suggests that biodiversity makes ecosystems more <u>stable</u> and more able to <u>resist</u> and <u>recover</u> from damage.

2) The more <u>plants</u> we have available, the more <u>resources</u> there are for developing <u>new food crops</u>.

3) Many new <u>medicines</u> have been discovered using chemicals produced by living things. For example, <u>digitalis</u>, a drug used to treat heart disease, was discovered in the <u>foxglove</u>. When a living organism becomes <u>extinct</u>, the unique chemicals it produces are no longer available.

Poor passenger pigeon — I blame Dick Dastardly...

I'm sure you already knew that letting things become extinct isn't good. What you might not have known though, is how it's often the fault of us <u>humans</u> that these species are gone for good. We now need to make sure we don't let anything else become extinct — we must think about <u>sustainable development</u>.

Revision Summary for Module B3

Well, I know I'm biased, but I thought that was a thoroughly interesting section. Evolution is quite amazing really, and always sparks a good few debates. Anyway, you know what time it is now. It's question time...

1) How many different species are there? A few? Hundreds? Thousands? Or millions?
2) How do fossils provide evidence that we evolved from simpler life forms?
3) How do scientists use DNA evidence to back up the theory of evolution?
4) What are the three main ideas of how life first appeared on Earth?
5) By which process do organisms adapt to a changing environment?
6) If a fox very nearly catches a rabbit and takes a big chunk out of its ear, will the rabbit's offspring be born with chunks out of their ears?
7) Give an example of a feature that has been artificially selected by humans in farm animals or crops.
8) Who came up with the idea of natural selection?
9) Why didn't some religious people agree with the theory of natural selection?
10) What non-religious reasons are there for people not believing the theory of natural selection?
11) Why was evolving a big brain a survival advantage for humans?
12) Do any human species other than Homo sapiens still exist today?
13) Who was Piltdown man? Did his evidence change the theory of human evolution?
14) List the five sense organs and say what receptors they contain.
15) What does CNS stand for? What does the CNS do?
16) Explain the difference between sensory and motor neurones.
17) Which travel faster through the body, nervous impulses or hormones?
18) Give one example of nervous communication and one example of hormonal communication.
19) List four factors that are essential for plants to live.
20) What is interdependence?
21) In a food web, if one organism dies out, will the other organisms be affected?
22) Give three reasons why a species might become extinct.
23) Give an example of a species that humans directly wiped out. How did they do this?
24) Give an example of a species that humans wiped out indirectly. Explain what happened.
25)*Why should we conserve biodiversity?

* Answer on page 100.

Organic and Intensive Farming

Thought farming was all much of a muchness? Well think again, because there are two distinctly <u>different</u> ways of producing food — <u>organic farming</u> and <u>intensive farming</u>.

Elements are Constantly Being Recycled...

There <u>isn't</u> a <u>never-ending</u> supply of the elements that living things need — luckily elements are <u>recycled</u>:

1) As plants grow they take in elements like <u>oxygen</u>, <u>nitrogen</u> and <u>carbon</u> through their leaves and roots.

2) When the plants <u>die</u> and <u>decompose</u>, most of these elements are returned to the <u>soil</u>. Others go into the <u>air</u> as gases like methane.

3) Some of the elements in plants become part of <u>animals</u> when the plants are eaten. These elements are also returned to the environment when the animals <u>poo</u> or when they <u>die</u> and <u>decompose</u>.

4) Dead animal and plant matter (and animal waste) is broken down by <u>microbes</u>. They convert it into compounds that are taken up by other plants and the <u>whole process starts again</u>.

...But Harvesting Crops Removes Them From the Soil

1) When a farmer removes crops from a field (at harvest time) some of the <u>elements</u> that the plants used to grow are <u>taken out</u> of the field for good, instead of being <u>returned</u> to the soil when the plants die and decay.

2) The most important element that's lost (and the only one you need to remember...) is <u>nitrogen</u>.

3) The elements lost from the soil need to be <u>replaced</u> — or the fertility of the soil will <u>decrease</u> and the next crop of plants won't grow properly.

4) How farmers replace the lost elements depends on whether they farm <u>organically</u> or <u>intensively</u>.

Different Farming Methods Replace Lost Elements Differently

① Organic Farming Relies on the Natural Recycling of Organic Matter

<u>Artificial fertilisers</u> are <u>banned</u> in organic farming, so <u>natural</u> substances and processes are used instead.

1) Organic farmers put <u>animal manure</u>, <u>compost</u> and <u>human sewage</u> onto their land as <u>fertilisers</u>. The human sewage is <u>heat-treated</u> first to destroy harmful microbes.

2) Organic farmers also grow "<u>green manure</u>". Plants are grown on fields and then <u>ploughed in</u> and left to rot.

3) Organic farmers also use <u>crop rotation</u> to help keep their soil fertile. They grow different crops each year in a <u>cycle</u> — for example, peas might be grown one year, cabbage the next and carrots in year three.

② Intensive Farming Relies on Artificial Fertilisers

1) Intensive farmers use man-made artificial <u>fertilisers</u> to put elements back into the soil.

2) Farmers can use <u>small volumes</u> of artificial fertilisers, because they contain much higher percentages of the elements the crops need than manure does.

3) Artificial fertilisers are spread on the <u>ground</u> as <u>pellets</u> or <u>sprayed</u> onto the <u>crops</u> as they grow.

So that's intensive farming — now do some intensive revision...

There are a few important things to take away from this page. Without a doubt the most important thing is that when you're eating <u>organic food</u>, you could be eating elements from someone else's <u>poo</u>.

Pest Control

So you've got all the <u>nutrients</u> your crop needs in the soil and everything looks <u>rosy</u> (especially if you're growing roses). But just when you thought nothing could go wrong, bam... a <u>plague of locusts</u>. Oh no.

There are Different Ways to Deal With Pests and Disease

<u>Pests</u> and <u>diseases</u> are an absolute nightmare for farmers because they can seriously <u>reduce crop yields</u>.

1) Some pests, like <u>aphids</u>, are insects that eat crop plants.

2) <u>Diseases</u> like <u>potato blight</u> can damage or kill crop plants.

Luckily, most pests and diseases can be <u>controlled</u>. The methods used depend on the type of farming.

(1) *Organic Farming Uses Natural Biological Processes*

<u>Organic farmers</u> aren't allowed to use man-made chemicals to deal with pests and diseases.

1) Pests can be controlled using <u>natural predators</u> — this is known as <u>biological control</u>. For example, ladybirds can be introduced into greenhouses to prey on greenfly pests.

2) <u>Crop rotation</u> (see previous page) is used to prevent the pests and disease causing organisms of one particular crop plant building up in an area.

3) Field edges are left <u>grassy</u> to encourage <u>larger insects</u> and other animals that feed on pests.

4) Varieties of plants that are best able to <u>resist</u> pests and diseases are chosen.

5) <u>Natural pesticides</u> can also be used. Some pesticides are completely natural, and as long as they're used <u>responsibly</u> they don't mess up the ecosystem.

(2) *Intensive Farming Relies on Chemicals*

Intensive farmers <u>spray</u> their crops with <u>man-made chemicals</u> to destroy pests and diseases.

1) Chemical pesticides are more <u>effective</u> than organic methods. They usually kill <u>all</u> of the pests and disease-causing organisms, which organic methods can't.

2) This means a <u>bigger yield</u> of crops with <u>fewer blemishes</u>.

3) But it's <u>not</u> all good — the spraying leaves a chemical <u>residue</u> on the crop. This could harm <u>humans</u> eating the plants, as well as the pests.

4) Chemical pesticides kill indiscriminately — this means that <u>not only pests</u> are killed but also other organisms that <u>could be beneficial</u>.

Organic Farmers Have to Follow Certain Rules

It's <u>illegal</u> to sell food as <u>organic</u> if it hasn't really been grown that way.

1) The UK government has set <u>national standards</u> that have to be met by organic farmers, e.g. concerning the use of chemicals.

2) The national rules ban the use of virtually all <u>artificial chemicals</u> and set standards for the way that pests and diseases are <u>controlled</u>. The levels of pesticides and other artificial chemicals in the <u>soil</u> have to be below a certain level before a farm can be classed as organic.

3) The standards for <u>meat</u> are just as strict. The animals must be allowed to move around freely, can only be fed on <u>organic</u> feed and can't be given <u>artificial hormones</u> to make them grow more quickly.

I'm arresting you on suspicion of carrot fraud...

<u>Both</u> types of farming have their own <u>costs</u> and <u>benefits</u> — for the <u>general public</u>, the <u>environment</u>, the <u>farmers</u>, etc. Organic farming is more <u>expensive</u> and you can't grow as much in one area, but it is more <u>sustainable</u> — it's less harmful to the environment and you don't have to manufacture chemicals.

Natural Polymers

When you think about <u>polymers</u> (if you ever do), you probably think of the <u>man-made</u> kind, like <u>plastics</u>. But as usual <u>Mother Nature</u> got there first. And <u>her</u> polymers don't choke seagulls and stuff either.

Carbohydrates and Proteins are Natural Polymers

Many of the chemicals found naturally in living things are <u>long-chain molecules</u> called <u>polymers</u> (see p.50).

1) Polymers are large molecules formed by combining lots of <u>smaller</u> molecules (called <u>monomers</u>) in a <u>regular pattern</u>.

2) <u>Complex carbohydrates</u> are polymers built by linking together <u>simple sugars</u> like <u>glucose</u>.

3) <u>Proteins</u> are huge polymer molecules built by linking small molecules called <u>amino acids</u>.

Carbohydrates Consist of Carbon, Hydrogen and Oxygen

<u>Carbohydrates</u> are a group of compounds that include <u>sugars</u> (the <u>monomers</u>) and the polymers <u>cellulose</u> and <u>starch</u>. They're all made up of the same elements — <u>carbon</u>, <u>hydrogen</u> and <u>oxygen</u>.

1) The name 'carbohydrate' comes from the fact that they're basically made of <u>carbon</u> and <u>water</u>.

2) The <u>simplest</u> carbohydrate is the sugar <u>glucose</u>, $C_6H_{12}O_6$. This is the sugar that plants make from carbon dioxide and water, by <u>photosynthesis</u> (see p.58).

3) Other carbohydrates like <u>starch</u> (for energy storage in plants) and <u>cellulose</u> are made by linking <u>glucose</u> molecules together in chains.

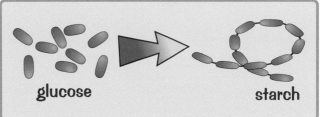

glucose   starch

Proteins Contain Carbon, Hydrogen, Oxygen and Nitrogen

<u>Proteins</u> are a huge group of compounds. They're all made from <u>amino acids</u> and consist mostly of <u>carbon</u>, <u>hydrogen</u>, <u>oxygen</u> and <u>nitrogen</u>.

1) Plants produce their own amino acids, which are then linked together into <u>long chains</u> to form <u>proteins</u>.

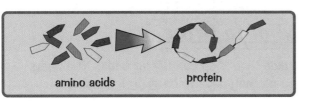

amino acids   protein

2) Animals have to get their amino acids from the <u>plants</u>. They break down plant proteins during <u>digestion</u> to give amino acids, and then join them back up in a <u>new order</u> in their cells to make <u>new proteins</u>.

3) Amino acids all have the <u>same basic structure</u> with <u>slight variations</u>.

4) There are only about <u>20</u> different amino acids used in the human body. It's the <u>order</u> in which they're connected up that makes each protein different.

Zzzzzzzzzzzzzzz... What? Who? No, of course I wasn't asleep...

It's pretty <u>mind-numbing</u> this, isn't it? Still, dull pages are bound to crop up from time to time and you've just got to grit your teeth and <u>learn</u> them. If you're still awake please learn the <u>elements</u> that <u>carbohydrates</u> and <u>proteins</u> are made from — that's the key thing you need to know.

Digestion

You get food from animals and plants. But you don't want their carbohydrates and proteins, you want your own human-type ones. So first you break down the polymers you eat, and then you build new ones.

Digestion Involves Breaking Down Big Molecules

After you've chewed your food up and your stomach's had its turn at churning it up even further, it's still made up of quite big molecules, namely: starch, proteins and fats.

These are insoluble and too big to diffuse into the blood, and so they are broken down in the small intestine into smaller, soluble molecules like glucose and amino acids. These can then move into the blood and be transported to all the cells of the body where they're needed.

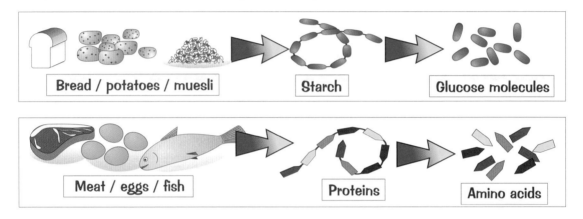

| Bread / potatoes / muesli | Starch | Glucose molecules |

| Meat / eggs / fish | Proteins | Amino acids |

Amino Acids Are Built Up into New Proteins

1) As cells grow they need new proteins, which they make from the amino acids flowing past in the blood.

2) Each protein has a different sequence of amino acids. One way to think of it is like the letters of the alphabet — every word (protein) is a different arrangement of the same letters (amino acids).

3) Most of your body is made from proteins. They're the main part of your skin, hair, muscles and tendons.

4) Haemoglobin (the red stuff in red blood cells) is a protein called globin attached to a substance called haem. Haem contains iron and is the substance that carries oxygen around your body.

Excess Amino Acids Are Disposed of in Urine

Animals often eat protein that contains more amino acids than they can use at once. The body can't usually store amino acids and use them later on, so they have to be excreted instead.

1) Excess amino acids in the blood are taken to the liver to be broken down.

2) They are converted into a soluble substance called urea which is then released into the blood.

3) As the blood passes through your kidneys, the dissolved urea is removed and passes out of the body in the urine.

Digestion? I'm sure they told me this was chemistry...

Ah you see, it's not like the old days when biology was pondweed, chemistry was Bunsen burners and physics was light bulbs. Now that you're older, science is all overlapping and full of issues. It's more relevant to real life, of course — but sometimes I still kind of miss the Bunsen burners.

Insulin and Diabetes

Everyone knows that it's <u>not</u> healthy to eat loads of <u>processed foods</u> and to be <u>overweight</u>, and yet loads of us still eat junk. That's why more people are developing <u>type 2 diabetes</u>, and at a younger age too.

Diabetes _is_ When The Body Can't Control Blood Sugar Levels

The body produces a hormone called insulin in the pancreas. Insulin helps glucose to move from the blood into body cells. This reduces blood glucose and gets the glucose where it needs to be to release its energy. Insulin also controls the storage of glucose. The amount of insulin you produce depends on your <u>blood sugar level</u>. The more sugar you've eaten and released into the blood, the more insulin is produced.

1) <u>Processed foods</u> often contain lots of sugar. For example, a can of cola contains 35 g of sugar (about 8 teaspoons). Sugar enters the bloodstream very <u>quickly</u>, making your blood sugar level <u>shoot up</u>. That's why it's better to eat <u>complex carbohydrates</u> (in brown bread, rice and cereals), which are broken down into sugar <u>gradually</u>.

2) <u>Diabetes</u> is an illness that occurs when the body <u>can't</u> control its blood sugar levels properly. There are two types of diabetes — <u>type 1</u> and <u>type 2</u>.

3) <u>Poor diet</u> and <u>obesity</u> increase your risk of developing <u>type 2</u> (but <u>not</u> type 1) diabetes.

The Two Types _of Diabetes Cause Problems in Different Ways_

The two types of diabetes <u>develop</u> for different reasons and are <u>treated</u> in different ways.

TYPE 1

1) Type 1 diabetes usually develops in <u>young</u> people and happens when the <u>pancreas</u> stops producing <u>insulin</u>, for reasons that aren't fully understood yet.

2) Because the body can't produce insulin, it can't remove <u>sugar</u> from the blood and store it. Blood sugar levels can become so <u>high</u> that they damage the body, possibly even causing coma and death.

3) Type 1 diabetes is treated by <u>daily injections of insulin</u> and by controlling the <u>diet</u> to help control the blood sugar levels.

TYPE 2

1) Type 2 diabetes usually affects <u>older</u> people and symptoms develop gradually. They include weight loss, needing to wee often and tiredness. It can develop if the body stops making enough <u>insulin</u> because it can't respond properly to high blood sugar levels. Or the body might be producing enough insulin, but stops <u>responding</u> to it normally. Both these things are linked to a <u>poor diet</u> and <u>obesity</u>.

2) Type 2 diabetes is controlled by improving the <u>diet</u>, <u>losing weight</u> and <u>exercising regularly</u>. Exercise can help to keep the blood sugar level stable and helps prevent obesity. Sometimes insulin is also taken.

3) Type 2 diabetes is becoming more and more common in <u>young</u> people. This is because of increasing obesity due to poor diet and lack of exercise.

Just 8 teaspoonfuls of sugar leads to obesity and diabetes...

One thing that they're keen on in science these days is <u>risk</u>, and this is just the kind of place they'd spring it on you. For example, what <u>risks</u> are associated with being <u>obese</u>? Why might people <u>accept</u> the risk involved in having a poor diet? (Think about stuff like <u>convenience</u>, <u>cost</u>, wide <u>availability</u> of processed foods.) You must be able to discuss <u>personal choices</u> in terms of <u>balancing</u> risks and benefits.

Harmful Chemicals in Food

You need to <u>eat</u> to live. But some foods can do you more <u>harm</u> than good...

Some Foods Are Dangerous Naturally...

Some plants contain <u>poisonous chemicals</u>, some are poisonous unless <u>cooked properly</u> and some cause <u>allergic reactions</u> in some people.

1) A lot of <u>mushrooms</u> are poisonous. Only a few are <u>deadly</u>, but many can cause <u>stomach upsets</u>.

2) <u>Cassava</u>, a plant found in South America, has a <u>floury root</u> that's widely eaten. In the root are compounds that produce lethal <u>cyanide</u> in the liver if eaten, but luckily it also contains a substance that <u>destroys</u> these compounds if it's <u>mixed</u> with them. This is normally done by either chopping up the roots and <u>boiling</u> them, or by <u>fermenting</u> them.

3) As many as one in 200 people are <u>allergic</u> to <u>proteins</u> found in <u>peanuts</u> and other nuts. In extreme cases this can be <u>fatal</u>, but more often the symptoms are a <u>rash</u> and <u>swelling</u> of the mouth and throat.

4) <u>Gluten</u>, a <u>protein</u> found in <u>wheat</u>, <u>rye</u> and <u>barley</u>, can cause <u>allergic reactions</u> in some people. Symptoms <u>vary</u> and may include rashes and swellings, stomach pain and vomiting, diarrhoea and bloating or even breathing problems. People with this allergy have to eat a <u>low gluten</u> or <u>gluten-free</u> <u>diet</u>.

...Others Contain Chemicals Left Over From Farming...

<u>Pesticides</u> and <u>herbicides</u> used in farming can be present in small amounts on <u>fruit</u> and <u>vegetables</u>.

1) These residues are usually in <u>very small amounts</u> and can be removed by careful <u>washing</u> and <u>peeling</u>.

2) The <u>long-term effect</u> on health of eating small amounts of these chemicals is not fully understood, so many people now do their best to avoid them. A recent study has linked residues with an increased risk of <u>Parkinson's disease</u> — a degenerative brain disease.

...Others Develop Dangerous Chemicals During Storage...

Some foods can contain harmful chemicals if they've not been <u>stored</u> properly.

1) <u>Nuts</u>, <u>cereals</u> and <u>dried fruit</u> can become contaminated by a poisonous substance called <u>aflatoxin</u>.

2) Aflatoxin is produced by a <u>mould</u> that can grow on these foods if they're not stored correctly.

3) Aflatoxin can cause <u>liver cancer</u>.

4) <u>Bird food</u> is a particular problem, as it doesn't have to meet the same <u>standards</u> as nuts and seeds for human consumption. Birds and other animals can be <u>killed</u> if they eat contaminated food.

...And Cooking Can Form Other Dangerous Chemicals

<u>Burning</u> foods can produce dangerous chemicals.

1) These chemicals are usually formed when food is cooked at <u>high temperatures</u> (over 150 °C or so).

2) They can be formed in a whole range of different foods, including meat and fish.

3) Such chemicals have been shown to cause <u>cancer</u> in animals by altering their <u>DNA</u>.

4) The amounts of these chemicals produced can be <u>reduced</u> by choosing <u>lean</u> meat or fish, not letting <u>flames</u> touch the food and by cooking it at a <u>lower temperature</u> for longer.

Think twice before you stick another shrimp on the barbie...

Here's that <u>risk and benefit</u> stuff again — scientists are <u>always</u> coming up with new ways that various foods are <u>bad</u> for us. But if you spend too much time analysing <u>all</u> those risks, you'd <u>never eat</u> at all.

Food Additives

You're not finished with food issues yet, I'm afraid. Here are all the chemicals that are put in on purpose.

Lots of Foods Contain Additives These Days

1) Food colours (should) make food look more appetising. They're often used in sweets and soft drinks. A place you might not expect to find food colouring is in mushy peas, which contain a green dye.

2) Flavourings, fairly obviously, are added to foods to give them a new taste — e.g. adding an orange flavour to a soft drink. They could be extracted from a natural substance or made artificially. Flavour enhancers are a bit different — they bring out the taste and smell of food without adding a taste of their own (they're not flavourings as such). They're often added to ready meals.

3) Artificial sweeteners (like saccharin and aspartame) are used in things like diet foods and drinks instead of sugar. They taste sweeter than sugar so you don't have to use as much, making the products lower in calories.

4) Antioxidants stop some foods from going off when they react with oxygen. Oxygen can turn the fat in food into nasty-smelling and nasty-tasting substances — e.g. butter goes rancid when it's exposed to the air. Antioxidants are added to foods that contain fat or oil, e.g. sausages, to stop them reacting with oxygen.

5) Preservatives are added to many foods to prevent the growth of harmful microbes. The food can then be stored for longer before it goes off. Look out for the common preservative, sodium benzoate.

Emulsifiers and Stabilisers Help Oils and Water Mix

1) Oil and water naturally separate into two layers with the oil floating on top of the water — they don't "want" to mix. Emulsifiers help to stop the two liquids in an emulsion separating out — you'll find them in things like mayonnaise.

2) Stabilisers are added to foods to help emulsions stay mixed and to thicken them. They're added to lots of foods, e.g. tomato sauce, ice cream and many other desserts.

The Use of Food Additives is Regulated

Food additives have to pass a safety test before they can be used.

1) In the EU (including the UK), all food additives have to pass a safety test and are then given an E number. Even oxygen has an E number, because it's used with gas-packed vegetables.

2) The standards set by the EU are different from those used in other countries. Some substances are allowed in the EU but not in other countries and vice versa.

3) Despite these safety tests, some additives are still thought to cause health problems in some people. For example:

- Some artificial food colourings have been linked with allergies and hyperactivity.
- Sulfur dioxide in dried fruit has been linked with asthma.
- Artificial sweeteners, e.g. aspartame, have been linked with hyperactivity and behavioural problems.

OK, enough — this is putting me off my chemical-filled lunch...

E numbers are something else that have had a very bad press lately. And fair enough, luminous coloured sweeties that have kids bouncing off the walls are probably a bit unnecessary. But some of the chemicals that are added are really useful, like preservatives that stop harmful microbes growing.

Keeping Food Safe

Reducing the risk of a hazard, e.g. a pesticide residue on food, costs money — it's expensive to monitor the levels, remove the chemical, etc. Governments have to balance the risks and costs to reach an acceptable level of risk.

Foods Can Be Made Safer but Not Risk-free

Everything you do carries a certain amount of risk, even something as simple as eating a meal. New technology based on scientific advances introduces new risks, and these have to be limited.

1) For example, scientists genetically modify some crops to give them new characteristics, like better yields and resistance to diseases. However, some people are worried that this new technology might not be safe, and so research into GM crops is strictly regulated. Food containing material from GM organisms must be clearly labelled so that consumers can avoid it if they want.

2) Scientists are also responsible for many of the chemicals sprayed on or added to foods (herbicides, pesticides, flavourings, preservatives etc.). These have their advantages, but they also pose risks (see p.74). Governments and other organisations evaluate these risks (see below).

3) No food can ever be guaranteed to be completely safe. There are too many stages in the food chain and too many different ways that a food could be contaminated. New allergies and food intolerances can develop at any point in a person's life. You simply have to accept a small amount of risk, and take what steps you can to make sure it's as low as possible.

4) For example, a lot can be done to reduce the potential risks of eating meat:

- The farmer should care for the animals properly.
- The animals should be slaughtered in a hygienic environment.
- The butcher must keep the meat in a clean and cold place.
- Members of the public should store and prepare the meat in the safest way.
- The government aims to ensure all this happens by licensing places like slaughterhouses, setting out guidelines, enforcing laws and educating the public.

Scientists Decide Safe Levels of Chemicals in Food

Foods are regulated to make sure they don't pose a significant risk to health.

1) Potentially dangerous chemicals get into food from a variety of sources and at every stage in the food supply chain — from pesticides in the fields, to processing, packaging and storage.

2) Scientific advisory committees, which aren't connected to the food industry, carry out risk assessments to help set safe limits for the levels of chemicals allowed in food.

3) The Food Standards Agency (FSA) is an independent food safety watchdog set up by an Act of Parliament to protect our health and consumer interests in relation to food.

4) The FSA offers advice to consumers and food producers on all aspects of food safety, labelling, diet, farming and hygiene. It also checks that legislation on these issues is being followed properly, for example by supporting food sampling programmes. These are carried out regularly by local authorities and involve testing various foods from different sources to make sure they're safe to eat.

I laugh in the face of danger — see, I haven't washed this pear...

I told you — these examiners love the whole idea of risk. In their spare time I expect they all go skydiving. This topic is an ideal way for them to introduce ideas about risk, but it's not the only place it could come up in an exam, so make sure you understand the general ideas behind these examples.

Eating Healthily

So the Government and various organisations are keeping a beady eye on the companies and individuals producing our food, making sure they don't <u>poison</u> us. But the Government can't be responsible for everything you put in your mouth.

Individual Choices Can Help Make Food Safer

Everyone should be aware of the <u>possible harmful effects</u> of the various <u>chemicals</u> in their food. Individuals who want to <u>reduce</u> their exposure to potentially harmful chemicals can take several steps:

1) Choose food produced in a way that <u>minimises</u> the amount of artificial chemicals applied to it. This usually means eating <u>organic</u> food.

2) <u>Wash</u> the food carefully or <u>peel</u> it before eating it.

3) <u>Store</u> and <u>cook</u> the food in the way recommended on the packaging.

People can also use the <u>labels</u> on food to find out more about the consequences for their health.

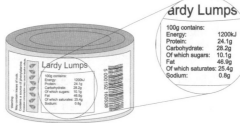

- Food labels give <u>detailed information</u> about things like the amounts of each <u>type of fat</u>, and sometimes whether this is high, medium or low compared with <u>other foods</u>.

The label should tell you:

- <u>How</u> and <u>where</u> the food was produced.

- What it <u>contains</u> and whether it contains substances that people might be <u>allergic</u> to.

Many labels also give you an idea of the <u>recommended daily amounts</u> of different substances you should be eating, and <u>how much</u> of that daily amount the product contains.

Choosing Food Isn't Just About What it Tastes Like

1) People <u>don't</u> all eat exactly the same diet. There's so much <u>choice</u> that two people's diets might be <u>completely different</u> — one person might avoid fruit and vegetables altogether, and the other might be a vegan who eats a diet of organic foods.

2) More and more people <u>are</u> becoming concerned about the food they eat and many are turning to <u>organic</u> food. Organic food is often seen as <u>safer</u> because fewer artificial chemicals are used in producing it, and some people see it as more <u>natural</u> and <u>nutritious</u>. For these people, the potential <u>risks</u> posed by eating food from intensive farms don't seem <u>worth</u> the benefits.

3) But there <u>are</u> benefits to intensively grown foods. They're <u>cheaper</u> (an important factor for many people). They often <u>look</u> more attractive, and there tends to be more <u>choice</u>.

4) <u>Processed ready meals</u> are <u>quick</u> and <u>convenient</u>. For some these benefits outweigh the potential <u>risks</u> posed by the <u>additives</u> in the food (such as high <u>sugar</u>, <u>fat</u> and <u>salt</u> contents).

5) All food you buy is subject to lots of government regulations, so unless you have an <u>allergy</u> to one of the ingredients, it's unlikely to <u>poison</u> you. This is enough for some people to dismiss all the risks of an unhealthy diet and eat whatever they like.

Your diet is a personal choice, and each person <u>balances</u> the <u>risks</u> and <u>benefits</u> of eating different foods for themselves based on their <u>own opinions</u>. Advice from the Government or their doctor, and what they've heard from other people and from the media might all <u>influence</u> their final decision.

It's the fast food giants versus Gillian McKeith...

Isn't it crazy that even though there's more choice these days, and more and more people are eating <u>free range</u>, <u>organic</u>, <u>local</u>, and whatever, <u>obesity</u> is still one of the biggest causes of death in countries like the USA and UK. That's personal choice for you. Put down that pie and get out on your bike.

Revision Summary for Module C3

Okay, I'm sure you know what to do by now — and if by some chance you've forgotten, that great big list of questions should give you a clue. Go through and try them all, making a note of any you can't do. Then go back through the section and find the answers to the ones you were stuck on. And I warn you, if you don't try these questions, whenever you try to grow rhubarb it will not sprout, and whenever you make custard it will turn out lumpy.

1) How is dead animal and plant matter turned into compounds that plants can use?

2) Give three ways that organic farmers can replace the nutrients that are lost from their soil.

3) Give two ways that an organic farmer can limit the number of diseases affecting his or her crops.

4) Describe the advantages and disadvantages of using chemical pesticides to kill pests.

5) What must a farmer who grows crops and produces animals for meat do for the food to be classed as organic?

6) Name two carbohydrate polymers.

7) Which element is found in proteins but not in carbohydrates?

8) Name the monomer molecules that make up a protein polymer.

9) Why can the starch in food not be absorbed into the blood straight away?

10) Name the protein attached to haem in red blood cells.

11) Where are excess amino acids broken down in the body? How are the products of this process excreted?

12) What normally happens to excess sugar in the body?

13) How is type 1 diabetes usually treated?

14) Suggest why type 2 diabetes is increasing in young people.

15) Explain why the South American plant, cassava, should not be eaten raw.

16) What is aflatoxin? What problem can it cause if eaten by humans?

17) Give a health problem that might be caused by eating burnt foods.

18) Why are the following added to food?   a) sodium benzoate   b) saccharin

19) Ice cream contains two different kinds of chemical to prevent it separating out — what are they?

20) Give two examples of problems that have been linked to additives passed as safe to use by the EU.

21)* Why can no food ever be guaranteed to be completely safe?

22) Explain how scientific advisory committees and the FSA limit the risks from food.

23) Suggest three ways that individuals can limit the risks from food.

24) What kinds of information can consumers get from food labels?

25)*Outline some of the risks and benefits of eating intensively produced vegetables.

* Answers on page 100.

Radioactivity

Nuclear radiation — yet another type of radiation. But you do need to read these next few pages. Sorry.

Atoms Consist of a Nucleus Plus Orbiting Electrons

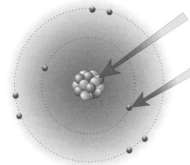

The nucleus contains protons and neutrons. It makes up most of the mass of the atom, but takes up virtually no space — it's tiny.

The electrons are really really small.

They whizz around the outside of the atom. Their paths take up a lot of space, giving the atom its overall size (though it's mostly empty space).

Radioactive Elements Emit Nuclear Radiation

1) Radioactive atoms are unstable — they break up (decay) to make themselves more stable.

2) Unstable atoms decay at random. If you have 1000 unstable atoms, you can't say when any one of them is going to decay, and you can't do anything at all to make a decay happen.

3) Each atom just decays quite spontaneously in its own good time. It's completely unaffected by physical conditions like temperature or by any sort of chemical bonding etc.

4) When an atom decays, it spits out one or more of the three types of radiation — alpha, beta, gamma.

ALPHA Radiation is Slow and Heavy

1) Alpha (α) particles are relatively big and heavy and fairly slow-moving.

2) So they don't penetrate far into materials — they're stopped quickly.

BETA Radiation is Lighter and More Penetrating

1) Beta (β) particles move quite fast and they are quite small.

2) They penetrate moderately into materials before they're stopped.

GAMMA Radiation is an Electromagnetic Wave

1) After spitting out an alpha or beta particle, the nucleus might need to get rid of some extra energy. It does this by emitting a gamma ray.

2) Gamma (γ) rays are a type of EM radiation (see p.54) — they have no mass.

3) They can penetrate a long way into materials before they're stopped.

Remember What Blocks the Three Types of Radiation...

Alpha particles are blocked by paper.
Beta particles are blocked by thin aluminium.
Gamma rays are blocked by thick lead.
Of course anything equivalent will also block them, e.g. skin will stop alpha but not the others, a thin sheet of any metal will stop beta, and very thick concrete will stop gamma just like lead does.

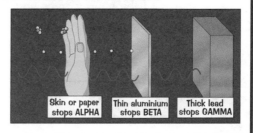

Skin or paper stops ALPHA   Thin aluminium stops BETA   Thick lead stops GAMMA

I once beta particle — it cried for ages...

So you can tell which kind of radiation you're dealing with by what blocks it. If it gets through paper, it could be either beta or gamma. If it gets through a sheet of aluminium, it must be gamma.

Half-Life and Background Radiation

Radioactivity is measured in <u>becquerels</u> (<u>Bq</u>) or counts per minute (cpm). 1 Bq is one <u>decay per second</u>.

The Radioactivity of a Sample Always Decreases Over Time

1) Each time an unstable nucleus <u>decays</u> and emits radiation, that means one more <u>radioactive nucleus</u> <u>isn't there</u> to decay later.

2) As more <u>unstable nuclei</u> decay, the <u>radioactivity</u> of the source <u>as a whole</u> <u>decreases</u> — so the <u>older</u> a radioactive source is, the <u>less radiation</u> it emits.

3) <u>How quickly</u> the activity <u>decreases</u> varies a lot. For <u>some</u> isotopes it takes <u>just a few hours</u> before nearly all the unstable nuclei have <u>decayed</u>. For others it can take <u>millions of years</u>.

4) The problem with trying to <u>measure</u> this is that <u>the activity never reaches zero</u>, which is why we have to use the idea of <u>half-life</u> to measure <u>how quickly the activity decreases</u>.

5) Learn this <u>important definition</u> of <u>half-life</u>:

> <u>HALF-LIFE</u> is the <u>TIME TAKEN</u> for <u>HALF</u> of the <u>radioactive atoms</u> now present to <u>DECAY</u>.

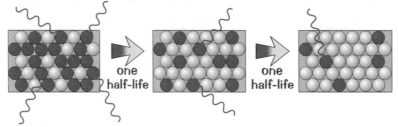

6) A <u>short half-life</u> means the <u>activity falls quickly</u>, because <u>lots</u> of the nuclei decay in a <u>short time</u>.

7) A <u>long half-life</u> means the activity <u>falls more slowly</u> because <u>most</u> of the nuclei don't decay <u>for a long time</u> — they just sit there, <u>basically unstable</u>, but kind of <u>biding their time</u>.

Background Radiation is Everywhere All the Time

We're constantly exposed to <u>very low levels</u> of radiation without us noticing — sneaky. It's called <u>background nuclear radiation</u>, and it's all around us all the time.

It comes from:

1) <u>NATURAL RADIOACTIVE SUBSTANCES</u> in the <u>air</u>, in <u>soil</u>, in <u>living things</u>, in the <u>rocks</u> under our feet...

2) <u>SPACE</u>: cosmic rays — these come mostly from the <u>Sun</u>.

3) <u>HUMAN ACTIVITY</u> — e.g. from <u>nuclear explosions</u> or <u>waste from nuclear power plants</u>, although this is usually a <u>tiny</u> proportion (<1%) of the total background radiation.

> <u>Radioactive sources</u> are considered to be "<u>safe</u>" when the radiation they are emitting is at about the <u>same level</u> as the <u>background radiation</u>. The <u>half-life</u> of the source gives an idea of how long it will take for this to happen.
>
> E.g. strontium-90 has a half-life of 29 years. A sample emitting 16 Bq will take <u>four half-lives</u>, 116 years, to reach roughly the background count of 1 Bq. (The activity of the sample <u>halves</u> after every half-life: 16 Bq → 8 Bq → 4 Bq → 2 Bq → 1 Bq.)

Half-life of a box of chocolates — about five minutes...

To measure half-life, you time how <u>long it takes</u> for the number of decays per second to halve — and this can vary from fractions of a second (you'd need to be mighty quick on the stopwatch) to thousands of millions of years (which is quite a while to wait with a Geiger counter and pencil).

Danger from Nuclear Radiation

Radioactive materials can be really useful, but they're also <u>dangerous</u> — they need <u>careful handling</u>.

Nuclear Radiation Causes Ionisation

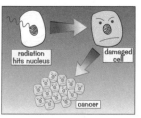

radiation hits nucleus

damaged cell

cancer

1) Alpha, beta and gamma radiation are all <u>ionising radiation</u> — they can <u>break up</u> molecules into smaller bits called <u>ions</u>.

2) In humans, ionisation can cause <u>serious damage</u> to the cells in the body.

3) A high dose of radiation tends to <u>kill cells</u> outright, causing <u>radiation sickness</u>. Lower doses tend to <u>damage cells</u> without killing them, which can cause <u>cancer</u>.

4) Which <u>type of radiation</u> is most <u>dangerous</u> depends on whether it's <u>inside</u> or <u>outside</u> the body. E.g. <u>alpha</u> sources are very dangerous <u>inside</u> the body, but they're relatively harmless <u>outside</u>.

5) Radioactive materials put people at risk through either:

    <u>IRRADIATION</u> — being exposed to radiation <u>without</u> coming into contact with the source. The damage to your body <u>stops</u> as soon as you leave the radioactive area.

    <u>CONTAMINATION</u> — <u>picking up</u> some radioactive material, e.g. by <u>breathing it in</u>, getting it on your skin or <u>drinking</u> contaminated water. You'll <u>still</u> be exposed to the radiation once you've <u>left</u> the radioactive area.

Sieverts _Show_ Possible Harm _from Nuclear Radiation_

1) How likely you are to <u>suffer damage</u> if you're exposed to nuclear radiation depends on the <u>radiation dose</u>. Radiation dose is measured in <u>sieverts</u> (<u>Sv</u>) or more usually <u>millisieverts</u> (<u>mSv</u>), and it takes into account the <u>type</u> and <u>amount of radiation</u> you've been exposed to, and the part of the body exposed.

2) The following data shows some <u>radiation doses</u> and their <u>effects</u> on the body:

2 mSv/year	Typical background radiation (see previous page) experienced by everyone.
9 mSv/year	Exposure by airline crew flying the New York to Tokyo polar route.
20 mSv/year	Current limit (averaged) for nuclear industry employees and uranium miners.
100 mSv/year	Lowest level at which an increase in cancer is clearly evident. Above this, the probability of cancer occurrence (rather than the severity) increases with dose.
1000 mSv/single dose	Causes (temporary) radiation sickness such as nausea and decreased white blood cell count, but not usually death. Above this, severity of illness increases with dose.
5000 mSv/single dose	Would kill about half of those receiving it within a month.
10 000 mSv/single dose	Fatal within a few weeks.

Researchers are still <u>arguing</u> about the effects of doses below 100 mSv/year, but most agree that a dose of 100 mSv per year for <u>10 years</u> <u>or more</u> will increase the risk of <u>cancer</u>.

Data from the World Nuclear Association

3) <u>Categories</u> of people who are at <u>higher risk</u> of radiation exposure include:

- <u>uranium</u> miners and processors
- workers in <u>nuclear power plants</u>
- <u>airline</u> staff (cosmic rays)
- <u>miners</u> (many rocks are naturally radioactive)
- some <u>medical</u> staff (e.g. radiographers)
- nuclear <u>researchers</u>

In Britain, people in high-risk industries have to <u>monitor</u> their radiation doses and have <u>regular check-ups</u>.

Revision sickness — never mind, only six pages to go...

Air hostesses — they should be wrapped head to toe in lead to keep the nasty radiation out, but because they wouldn't look as glam (and it'd cost the airlines too much money), they just get a silly hat.

Uses and Risks of Nuclear Radiation

Nuclear Radiation Can be Very Useful For...

...Treating Cancer

1) Since high doses of gamma rays will kill all living cells, they can be used to treat cancers.

2) The gamma rays have to be directed carefully and at just the right dosage so as to kill the cancer cells without damaging too many normal cells.

3) However, a fair bit of damage is inevitably done to normal cells, which makes the patient feel very ill. But if the cancer is successfully killed off in the end, then most patients feel it's worth it.

...Sterilising Medical Equipment

1) Gamma rays are used to sterilise medical instruments by killing all the microbes.

2) This is better than trying to boil plastic instruments, which might be damaged by high temperatures. You need to use a strongly radioactive source that has a long half-life, so that it doesn't need replacing too often.

...Sterilising Food

1) Food can be sterilised in the same way as medical instruments — again killing all the microbes.

2) This keeps the food fresh for longer, without having to freeze it or cook it or preserve it some other way.

3) The food is not radioactive afterwards, so it's perfectly safe to eat.

Using Radioactive Materials Brings Both Risks and Benefits

Nothing in life is completely safe. Scientific developments always bring new risks, e.g. new medical treatments can have nasty side effects, nuclear power stations might leak radioactive material. Like with any risky activity, using radioactive materials has benefits (and risks) for different groups of people. E.g. in building and running a nuclear power station (p.88):

1) Construction companies benefit from years of work in building the power station.

2) Local people benefit from new jobs — but might suffer from higher radiation exposure.

3) The national population benefits from a reliable energy source — but is at risk from nuclear waste.

4) There's a global benefit because nuclear power contributes a lot less to climate change than burning fossil fuels — but there's a global risk too, of a major accident like the Chernobyl disaster.

We're More Willing to Take Some Risks Than Others

1) Most of us are happy to take a risk if we know we're getting a substantial benefit. E.g. to help diagnose a medical condition, a patient might have radioactive material injected into their body as a tracer. This increases the patient's chances of developing cancer (very slightly) — but getting an accurate diagnosis (and then treatment) is a big benefit.

2) But there are some risks that you're not allowed to take — whatever the benefits. There are strict laws that set down exactly what you can and can't do with radioactive materials.

Gamma radiation — just what the doctor ordered...

Radiation isn't all bad, then. But (like just about everything else in life) there are risks involved. So the question you've got to ask yourself is this... do you feel lucky? Well do ya?   Ahem.

Electricity

You might be wondering what electricity has to do with radioactivity... bear with me.

Electricity is a Convenient Way to Supply Energy

1) Electricity is a <u>secondary energy source</u> because it is produced using <u>other</u> energy resources, e.g. by burning coal, nuclear reactions.

2) Electricity is <u>convenient</u> because it can be easily <u>transmitted</u> over long distances via the National Grid, and can be used in <u>many different ways</u>.

3) Power stations are used to generate electricity in <u>three stages</u>:

> 1. Energy is <u>released</u> from the <u>fuel</u> (usually by burning) and used to generate <u>steam</u>. → 2. The steam turns a <u>turbine</u>. → 3. A <u>generator</u> converts the movement of the turbine (kinetic energy) into <u>electricity</u>.

Supplying Electricity isn't Close to 100% Efficient

1) Unfortunately, most power stations <u>aren't very efficient</u>. They produce lots of <u>waste energy</u> (heat and noise) as well as useful electricity. Some energy is also lost as heat in the transmission wires as the electricity is <u>distributed</u> from the power station to people's homes.

2) <u>Sankey diagrams</u> let you see at a glance how much of the input energy is being <u>usefully employed</u>, and how much wasted.

3) The <u>thicker the arrow</u>, the more <u>energy</u> it represents — so you see a big thick arrow going in, then several smaller arrows going off it to show the different energy transformations.

4) The <u>width</u> of each arrow is <u>proportional</u> to the amount of energy it represents (measured in joules in this case).

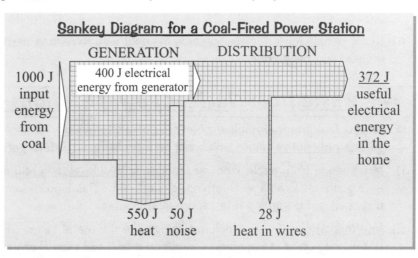

Sankey Diagram for a Coal-Fired Power Station

GENERATION   DISTRIBUTION

1000 J input energy from coal

400 J electrical energy from generator

<u>372 J</u> useful electrical energy in the home

550 J heat   50 J noise   28 J heat in wires

5) You can calculate the <u>efficiency</u> of the energy transfers shown in the diagram using this equation:

$$\text{Efficiency} = \frac{\text{USEFUL Energy OUTPUT}}{\text{Energy INPUT}}$$

Efficiency of <u>GENERATION</u>: $\text{Efficiency} = \dfrac{\text{electrical energy from generator}}{\text{energy input from coal}} = \dfrac{400}{1000} = \underline{0.4}$

Efficiency of <u>DISTRIBUTION</u>: $\text{Efficiency} = \dfrac{\text{useful electrical energy output}}{\text{electrical energy from generator}} = \dfrac{372}{400} = \underline{0.93}$

<u>OVERALL</u> efficiency: $\text{Efficiency} = \dfrac{\text{useful electrical energy output}}{\text{energy input from coal}} = \dfrac{372}{1000} = \underline{0.372}$

Skankey diagrams — to represent the smelliness of your socks...

The thing about loss of energy is it's <u>always the same</u> — whatever the process, some energy always disappears as <u>heat</u> and sound, and even the sound ends up as heat pretty quickly.

Generating Electricity — Non-Renewables

Fossil fuels and nuclear fuels are both non-renewable resources — at some point they're going to run out.

Fossil-Fuel Power Stations Release Carbon Dioxide — Not Ideal

1) At the moment, nearly three quarters of all the electricity generated in the UK comes from fossil fuels (oil, gas and coal) — which all contain carbon.

2) When fossil fuels are burned in power stations, the carbon is converted into carbon dioxide.

3) This carbon dioxide is released into the atmosphere and contributes to the greenhouse effect and climate change (see p.59-61).

4) Governments are trying to reduce carbon dioxide emissions by using alternative power sources. One possibility is nuclear power...

Nuclear Power Stations Release Energy by Splitting Atoms

1) A nuclear fuel, e.g. uranium, releases large amounts of energy when its nuclei undergo changes.

2) In nuclear power stations, uranium nuclei are split to release energy.

3) This process can be used to generate lots of electricity without releasing lots of carbon dioxide into the atmosphere (although mining and processing the fuel releases quite a bit).

4) Some people think nuclear power is the best way to reduce carbon dioxide emissions, but others think it's just too dangerous.

The Waste from Nuclear Power Stations is Hard to Deal With

Some people argue that nuclear power is not a sustainable technology — because it produces dangerous radioactive waste which will be a problem for future generations.

1) Most waste from nuclear power stations is 'low level' (slightly radioactive) — e.g. things like paper, clothing, gloves, etc. This kind of waste can be disposed of by burying it in secure landfill sites.

2) Intermediate level waste includes things like the metal cases of used fuel rods. It's usually quite radioactive — and some of it will stay that way for tens of thousands of years. It's often sealed into concrete blocks then put in steel canisters for storage.

3) High level waste is so radioactive that it generates a lot of heat. This waste is sealed up in glass and steel, then cooled for about 50 years before it can be moved to more permanent storage.

4) The canisters of intermediate and high level wastes could then be buried deep underground. However, it's difficult to find suitable places. The site has to be geologically stable (e.g. not suffer earthquakes), since big movements in the rock could break the canisters and radioactive material could leak out.

5) Even when geologists do find suitable sites, people who live nearby often object. So, at the moment, most intermediate and high level waste is kept 'on-site' at nuclear power stations.

6) There are very strict regulations about how radioactive waste is disposed of.

7) But the rules could change as we find out more about the dangers of radiation, and the pros and cons of storing waste in different ways. What's allowed now might be considered too risky in the future.

Nuclear power — people tend to get steamed up about it...

Building lots of nuclear power stations would mean we wouldn't have to rely on fossil fuels — and we wouldn't have to buy so much fuel from other countries. But do the risks outweigh the benefits...

Generating Electricity — Renewables

Most UK electricity comes from fossil-fuel power stations that burn either coal, oil or natural gas. The Government target is to have <u>10%</u> of our electricity generated from <u>renewable resources</u> by 2010.

Renewable Energy Resources are an Alternative to Fossil Fuels

1) Renewable resources include <u>wind</u>, <u>solar</u>, <u>biomass</u>, <u>wave</u>, <u>tidal</u>, <u>hydroelectric</u> and <u>geothermal</u> energy.

2) These will <u>never run out</u>.

3) If they damage the environment, they tend to do it in <u>less nasty ways</u> than fossil fuels.

4) The trouble is they <u>don't yet provide much energy</u> and some of them are <u>unreliable</u> because they depend on the weather.

5) Generating electricity from renewable energy resources is an example of <u>sustainable development</u> — it doesn't cause damage which harms the ability of future generations to meet their needs.

You need to know <u>two examples</u> of renewable energy resources, so here are a few more details about wind power and solar power:

Wind Power — Lots of Little Wind Turbines

1) Wind turbines have <u>blades</u>, a bit like a <u>windmill</u>.

2) Each wind turbine has its own <u>generator</u> inside it. The electricity is generated <u>directly</u> from the wind turning the <u>blades</u>, which <u>turn the generator</u>.

3) They need to be sited where it's <u>windy</u>, such as on a <u>hill</u>, near the <u>coast</u> or <u>offshore</u>. They need wind speeds of at least 3-4 m/s to work and have to be shut down if conditions get too stormy.

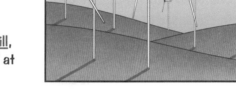

4) Their output is quite <u>variable</u> — and when there's <u>no wind</u>, there's <u>no power</u> at all.

5) But there are <u>no carbon dioxide</u> emissions, and <u>no fuel costs</u> (wind is free).

6) At the moment wind energy contributes about <u>1%</u> to the UK's overall electricity generation. This is likely to increase over the next 10 years to help meet targets on reducing carbon dioxide emissions.

Solar Power — Electricity from the Sun

1) Solar cells use special materials like <u>silicon</u> which can convert light energy into electricity <u>directly</u>.

2) They're often the best source of energy for appliances that don't use much energy (e.g. calculators), and in <u>remote areas</u> where there aren't many other options.

3) Solar energy can only be generated <u>during the day</u>, and ideally you need <u>plenty of sunshine</u>. Some of the electricity can be used to <u>charge a battery</u> for use during the night.

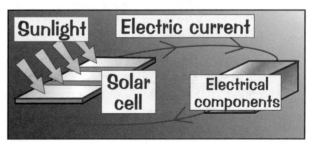

4) As with wind power, there are <u>no carbon dioxide</u> emissions, and <u>no fuel costs</u>.

Solar cells are like fried eggs — best sunny side up...

If you have your own solar panel or wind turbine, you can <u>sell back</u> any surplus electricity to the National Grid. So if you don't use much electricity but you <u>generate</u> a lot of it, you can actually make money instead of spending it. Nice trick if you can do it. Shame solar panels cost an arm and a leg...

Electricity in the Future

No Energy Source is Perfect

The table below shows comparison data for some large, modern power stations in the UK:

	Coal	Natural gas	Nuclear	Wind
Efficiency	36%	60%	38%	60%
Energy output per year (millions of units)	8000	5000	7000	150
CO_2 emissions per unit of electricity (g)	920	440	110	none
Cost of energy per unit	2.5-4.5p	2-3p	4-7p	3-4p

Talking about the efficiency of a wind farm doesn't mean all that much, though, since the 'fuel' is free and renewable.

The CO_2 emissions figures include the CO_2 released by mining and transporting the fuel, as well as burning it.

We Have to Look for the Best Compromise

We Need a Steady, Reliable Fuel Supply

1) All our uranium and over half of the coal we burn is currently imported from other countries. This makes us dependent on other countries and means transport costs and transport-related CO_2 emissions need to be considered.

2) Our supplies of coal, oil and gas will probably run out within 50 years or so, and uranium will run out eventually.

3) Renewable fuel sources such as wind, tides, solar and water are free and won't run out. But some of them are quite variable — it's not always sunny and the wind doesn't blow all the time.

We Need Enough Electricity... Obviously

The UK uses a lot of electricity. Renewable resources struggle to produce the same sort of energy output as conventional power stations.

We Need to Keep the Environmental Impact Down

1) Reducing carbon dioxide emissions (and other pollutants from fossil fuels) is a high priority.

2) Some energy resources directly affect wildlife habitats more than others. E.g. tidal barrages and most hydroelectric power (HEP) schemes work by damming water — this floods the land, destroying habitats. Wind farms take up a lot more room than other types of power station, but the land in between the turbines can be used.

3) Nuclear energy has different environmental impacts from other fuels. The mining of uranium leaves a lot of waste material, including radioactive rock and toxic metals. The decommissioning of a nuclear power station takes about 25 years, and how to get rid of nuclear waste hasn't been resolved yet.

4) Noise pollution is another consideration. Wind farms can be quite noisy close up, so some people don't want them near their homes. Coal-fired plants need to be supplied with coal, which can mean heavy traffic and the noise that goes with it.

5) And visual pollution. Conventional power stations are generally considered unattractive. Some people also find wind turbines ugly, and the best sites for HEP are often in areas of natural beauty.

Of course the biggest problem is that we use too much electricity...

It would be lovely if we could get rid of all the nasty polluting power stations and replace them with clean, green fuel... but it's not quite that simple. Renewable energy has its own problems too, and probably couldn't power the whole country without having a wind farm in everyone's backyard.

Module P3 — Radioactive Materials

Revision Summary for Module P3

Phew... what a relief — you made it to the end of yet another section. But don't run off to put the kettle on just yet — make sure that you really know your stuff with these revision questions.

1) What are the three types of particle found in an atom?
2) What do we mean by an 'unstable' atom?
3) Radioactive decay is spontaneous. Explain what this means.
4) Describe the properties of the three types of radiation: α, β and γ.
5) What substances could be used to block: a) α-radiation,   b) β-radiation,   c) γ-radiation?
6) Define half-life.
7) Give three sources of background nuclear radiation.
8) Describe what kind of damage radiation causes to body cells. What are the effects of high doses? What are the effects of lower doses?
9) What units are doses of radiation measured in? What three things does 'dose' take into account?
10) Give four categories of people who are at a higher than normal risk of exposure to radiation.
11) Describe how radioactive sources are used in each of the following:

    a) treating cancer,   b) sterilising medical equipment,   c) sterilising food.
12) Outline some arguments for and against the building of a nuclear power plant.
13) Suggest a reason why someone might be willing to expose themselves to nuclear radiation, despite the known risks associated with it.
14) What is a secondary energy source? Give an example.
15) Describe the three stages by which power stations generate electricity.
16)* The following Sankey diagram shows how energy is converted in a coal-fired power station.

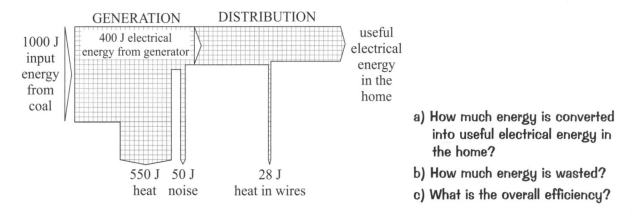

a) How much energy is converted into useful electrical energy in the home?

b) How much energy is wasted?

c) What is the overall efficiency?

17) Describe how nuclear power stations generate electricity using uranium fuel.
18) What are the main environmental problems associated with nuclear power?
19) Name three types of renewable energy resource.
20) Give one advantage and one disadvantage of using: a) wind power,   b) solar power.

* Answers on page 100.

Dealing with Tricky Questions

So, you've made it to the end of the book — well done. I'm afraid that's not quite it though...
you've still got the small matter of your exams to sort out.

The Exams for Units 1, 2 and 3 are Fairly Straightforward

Most of the questions you'll get in the exam will be pretty straightforward, as long as you've
learnt your stuff. But there are a few things you need to be aware of that could catch you out...

A Lot of Questions will be Based on Real-Life Situations...

...so learn to live with it.

All they do is ask you how the science you've learnt fits into the real world — that's all.

That equation you learnt still works... and that graph is still the same shape.

Don't worry that the theory you've learnt might not apply in different situations. IT WILL. For example...

> 4. European adders are found in a range of different colours from light brown to black.
>
> What are the possible causes of this variation?

It doesn't matter if you don't know anything about adders (who does?).
You just need to apply the stuff you've learnt about the genetic and environmental causes of variation.

There are Loads of Tick-the-Box Questions

The examiners seem fairly keen on questions where you have to choose from options by ticking boxes.

First things first — is it 'TICK' or 'TICKS'?
Check whether you can tick more than one box...

Then look through your options and use what you know to rule out the wrong 'uns.

You're only looking for one right answer.

One of the big things about chemical reactions is that none of the atoms just disappear so this sounds unlikely.

As I'm sure you remember from C1 (see p.18 for a quick reminder) this isn't true. The products of a chemical reaction don't have the same properties as the reactants.

This one's pretty much the opposite of the first and fourth ones — and since we've decided they're wrong, this one seems like a pretty good bet.

This one is saying something similar to the first one — that the total number of atoms changes during the reaction. We've already ruled that out.

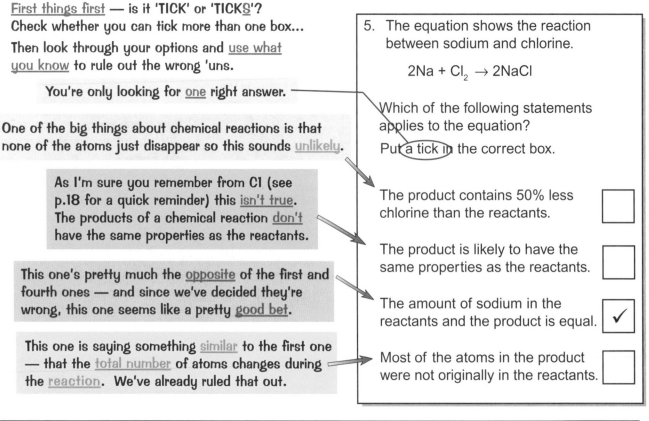

5. The equation shows the reaction between sodium and chlorine.

$$2Na + Cl_2 \rightarrow 2NaCl$$

Which of the following statements applies to the equation?
Put a tick in the correct box.

The product contains 50% less chlorine than the reactants. ☐

The product is likely to have the same properties as the reactants. ☐

The amount of sodium in the reactants and the product is equal. ✓

Most of the atoms in the product were not originally in the reactants. ☐

If at first you don't succeed — come back to it later...

It's true — you might get a stinker in the exam. So, do the straightforward ones first and then come
back to the tricky ones when you've done your best with the rest of the paper.*

Extract Questions

There might be a question on the exam where you have to answer questions about a <u>chunk of text</u>. If there is, here's what to do...

Questions with Extracts can be Tricky

1) Nowadays, the examiners want you to be able to <u>apply</u> your scientific knowledge to situations you've <u>never seen</u> before. Eeek.

2) The trick is <u>not</u> to <u>panic</u>. They're <u>not</u> expecting you to show Einstein-like levels of scientific insight (not usually, anyway).

3) They're just expecting you to use the science you <u>know</u> in an <u>unfamiliar setting</u> — and usually they'll give you some <u>extra info</u> too that you should use in your answer.

So to give you an idea of what to expect come exam time, use the new <u>CGP Exam Simulator</u> (below). Read the article and have a go at the questions. It's <u>guaranteed</u> to be just as much fun as the real thing.

Underlining or making notes of the main bits as you read is a good idea.

1. Blood glucose levels controlled by insulin.

2. Insulin added
   → liver removes glucose.

3. Not enough insulin
   → high blood glucose
   → death?

4. Two key methods involved in managing the problem

5. New areas of research - hopes for new treatments

This question is similar to the one on the opposite page — the best approach is to eliminate the definite 'nos' and then see what you have left. The article should give you some clues.

For this one you need to apply the arguments for and against embryonic research to the situation in the article.

All cells need energy to function, and this energy is supplied by glucose carried in the blood. The level of glucose in the blood is controlled by the hormone <u>insulin</u> — if the <u>blood glucose level gets too high, insulin is</u> introduced into the bloodstream by the pancreas, which in turn makes the <u>liver</u> remove glucose from the blood.

Diabetes (type I) is where <u>not enough insulin is produced</u>, meaning that a person's <u>blood glucose</u> level <u>can rise to a level that can kill them</u>. The problem can be controlled using a combination of these methods:

a) Avoiding foods <u>rich in carbohydrates</u>. It can also be helpful to take <u>exercise</u> after eating carbohydrates.

b) <u>Injecting</u> insulin before meals.

At a recent press conference, Dave Edwards from InsulinProducts plc said, "We have invested heavily in <u>embryonic stem cell</u> research and expect to be launching the final phase of our <u>testing into replacing</u> pancreas cells over the next year. We're <u>confident</u> that treatments for <u>diabetes</u> will change dramatically within the next decade. We believe that our research will improve the lives of the many diabetics who suffer daily, and further our scientific understanding".

(a) People with diabetes should eat sensibly after injecting insulin. Put a tick in a box to show the **best** explanation of this.*

The insulin will cause their glucose levels to increase too rapidly. ☐

The insulin will remove glucose from their blood — if no food is eaten the blood sugar level may drop too low. ☐

The insulin could damage their liver. ☐

The insulin will remove salt from their blood — if no food is eaten the salt levels could fall too much. ☐

(b) Put a ring around the sentence that provides a justification for research using embryos.*

Thinking in an exam — it's not like the old days...

*Answers on page 100.

It's scary when they expect you to <u>think</u> in the exam. But questions like this often have some of the answers <u>hidden</u> in the text, which is always a bonus. Just make sure you read <u>carefully</u> and take your <u>time</u>.

Interpreting Data

Chances are there'll be a few <u>data-related</u> questions of some kind in your exam — graphs, tables, that kind of thing. Here's an example to show you what you need to be able to do.

Get as Much Practice as You Can at Reading Graphs

The idea of graphs is to make <u>data</u> easier to <u>interpret</u>. And that's <u>generally true</u> — but they can be tricky.

So try to get <u>lots of practice</u> at reading graphs before the exam. And as I've said before, if you get one in the exam, <u>don't panic</u>.

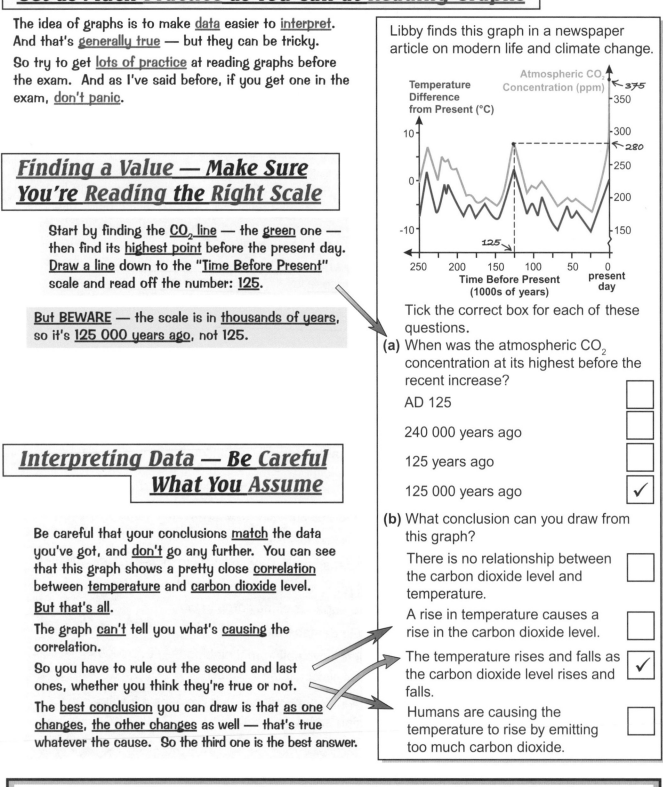

Libby finds this graph in a newspaper article on modern life and climate change.

Finding a Value — Make Sure You're Reading the Right Scale

Start by finding the <u>CO_2 line</u> — the <u>green</u> one — then find its <u>highest point</u> before the present day. <u>Draw a line</u> down to the "<u>Time Before Present</u>" scale and read off the number: <u>125</u>.

<u>But BEWARE</u> — the scale is in <u>thousands of years</u>, so it's <u>125 000 years ago</u>, not 125.

Tick the correct box for each of these questions.

(a) When was the atmospheric CO_2 concentration at its highest before the recent increase?

AD 125 ☐

240 000 years ago ☐

125 years ago ☐

125 000 years ago ☑

Interpreting Data — Be Careful What You Assume

Be careful that your conclusions <u>match</u> the data you've got, and <u>don't</u> go any further. You can see that this graph shows a pretty close <u>correlation</u> between <u>temperature</u> and <u>carbon dioxide</u> level.

<u>But that's all.</u>

The graph <u>can't</u> tell you what's <u>causing</u> the correlation.

So you have to rule out the second and last ones, whether you think they're true or not.

The <u>best conclusion</u> you can draw is that <u>as one changes, the other changes</u> as well — that's true whatever the cause. So the third one is the best answer.

(b) What conclusion can you draw from this graph?

There is no relationship between the carbon dioxide level and temperature. ☐

A rise in temperature causes a rise in the carbon dioxide level. ☐

The temperature rises and falls as the carbon dioxide level rises and falls. ☑

Humans are causing the temperature to rise by emitting too much carbon dioxide. ☐

Graphs are like MPs — they don't give you the whole picture...

There's not too much argument about the 'more CO_2 is making us warmer' theory now (although there <u>was</u> plenty of doubt previously, when the data was dodgy). But some people do still question whether the warming is <u>our fault</u> or just down to <u>natural variability</u> in the climate. Nothing's ever simple. Sigh.

Exam Skills — Paper Four

You've got to do an exam that's a bit out of the ordinary — Paper Four, Ideas in Context. As I've said a few times over the last few pages <u>DON'T WORRY</u> about it — just calmly read through this page and then <u>DO EXACTLY WHAT I SAY</u>.

You'll be Given Some Material <u>in Advance</u>

Before the exam you'll be given a <u>booklet</u> containing some <u>articles</u> that you'll be <u>examined</u> on. <u>Don't just</u> stuff it in the bottom of your bag with your PE kit — start working on it straight away.

The articles could be on <u>any science topic</u> that's <u>related</u> to the material on the specification. They could well be about something that's been in the news recently, or you could get an article about how <u>scientific understanding</u> has developed <u>over time</u>.

So, don't be surprised if some of them seem a bit <u>wacky</u> — they <u>will</u> be related to the stuff you've learned — it might just take you a while to figure out how...

Start work as <u>soon</u> as you get the booklet

1) Read all the articles <u>carefully</u> and <u>slowly</u> — take your time and make sure you <u>understand everything</u>.

2) <u>Look up</u> any words that you don't know.

3) If there are any <u>graphs</u>, <u>tables</u> or <u>figures</u> in the articles then study them carefully. Identify any <u>trends</u> and make sure you know what they <u>show</u> (see opposite for more on this).

4) You <u>don't</u> need to do any <u>extra research</u> on the topics but if you're struggling with the material then a bit of <u>extra reading</u> might help you to understand it — try textbooks and internet searches. <u>Don't</u> get too carried away though — you shouldn't need any extra knowledge to answer the questions.

5) Although you don't need to do research you <u>do</u> need to make sure you've revised the topics that the articles are about. So if there's one about <u>cystic fibrosis</u> make sure you've revised <u>genes</u>, <u>genetic inheritance</u> and all the stuff about <u>gene therapy</u>.

6) Remember to do all of the above for <u>all</u> of the articles — you'll have to answer questions on <u>all</u> of them in the exam.

7) It's a good idea to <u>highlight</u> important things in the booklet and make some <u>notes</u> as you go along, but remember that you can't take the booklet into the exam.

There'll be a <u>mixture</u> of questions in the exam

When you get to the exam you'll be given another copy of the articles, and some questions to go with them.

1) For some of the questions you'll need to <u>extract</u> information from the articles (see page 93).

2) Other questions will be about <u>analysing data</u> or <u>information</u> in the articles (see previous page).

3) Other questions will ask you about <u>related</u> topics from the specification — but you won't be expected to know anything that <u>isn't</u> on the specification or in the article.

I'd prefer an exam where you get the questions in advance...

The trick with this paper is to use the time before the exam to make sure you're really <u>comfortable</u> with the topics that are covered in the articles. If you're the type who watches the news and stuff then chances are you'll be familiar with some of the issues before you even get the booklet.

Index

Index

Index

Index

Answers

Revision Summary for Module B1 (page 16)

13)a) 100%

b) 0%

Revision Summary for Module P1 (page 35)

9) Currently, we don't understand well enough what 'warning signs' to look for to be able to say, reliably, when and where a jolt will happen. Strains in rocks only suggest an earthquake is more likely, not that it is certain.

21) The 'scientific community' (all the scientists working in the same field).

'Peer review' means that when a scientist has done some research and written a report about it, another scientist working in the same field reads the report before it's published. This 'reviewer' checks that the experiments have been conducted properly, written up in a detailed and unbiased way, and don't have any obvious flaws.

Revision Summary for Module B2 (page 45)

6) The inactive microbes still carry antigens which your immune system responds to — so your white blood cells produce antibodies to attack them. If you're infected with the same disease, these white blood cells can reproduce rapidly and release lots of antibodies to kill off the microorganisms before you become ill.

Revision Summary for Module C2 (page 53)

9 a) <u>14.2</u> (the anomalous result)

b) 8.1 + 8.3 + 8.1 + 8.0 + 8.3 + 8.4 + 8.0 + 8.2 = 65.4

$65.4 \div 8 = 8.175 = \underline{8.2 \text{ g/cm}^3}$ (to 1 d.p.)

12) Any three from: non-toxic, stiff, non-brittle, hard, fairly high melting point.

Revision Summary for Module P2 (page 62)

23) There could be health risks from heating tissues, e.g. in the brain. The amount of heating depends on the intensity of the radiation and the exposure time. Mobile phones emit fairly low intensity radiation, but are held close to the head — the intensity doesn't decrease much over this short distance. If you use a phone for a long time, you're exposed to more radiation, increasing the heating effect.

Revision Summary for Module B3 (page 72)

25) Without biodiversity, ecosystems would become unstable and less able to resist, and recover from, damage. Biodiversity is also important in the search for new medicines and new food crops.

Revision Summary for Module C3 (page 82)

21) You can't predict with absolute certainty your body's reaction to any food. Also, there are many ways that the food could have become accidentally contaminated (e.g. with pesticides and herbicides). And high temperature cooking might lead to health problems.

25) The risks are mainly from the chemicals used in the production of these foods. By law, residues of pesticides, etc. should only exist in small amounts, but nobody is certain of the long-term effect of eating small amounts repeatedly. Some residues have been linked with diseases e.g. Parkinson's.

The benefits include price, choice and convenience. Some people also prefer the unblemished appearance and uniformity of intensively farmed foods.

Revision Summary for Module P3 (page 91)

16a) 1000 − (550 + 50 + 28) = <u>372 J</u>

b) 550 + 50 + 28 = <u>628 J</u>

c) 372 ÷ 1000 = <u>0.372</u>

Answers to Extract Questions (page 93)

a) The insulin will remove glucose from their blood — if no food is eaten the blood sugar level may drop too low.

b) The sentence "We believe that our research will improve the lives of the many diabetics who suffer daily, and further our scientific understanding." should be ringed.